INTEGRATED FISH FARMING

ABOUT THE BOOK

This book on *Integrated Fish Farming* is an attempt to deal with all aspects of integrated aquaculture practice by taking as a case study of pig-fish farming.It would serve as a reference book for researchers and fisheries students and as guide for planners and development personnels, as the compendium for the enterpreneursand even the fish farmers.the book will bring a desirable change in the production system thereby providing better economic returns for all those connected with the aquaculture system including exhibiting a socio-economic upliftment of the poor fishermen.

THE AUTHORS

Dr. Nityanand Pandey, BFSc., MFSc., Ph.D. is Assistant Professor in the Department of Aquaculture, College of Fisheries, G.B. Pant University of Agriculture & Technology, Pantnagar, U.S. Nagar (Uttaranchal), India. Dr. Pandey is specialized in Aquaculture and several research papers published in national and International journals. Dr. Pandey has several extension booklet and popular articles in his credit and is a fellow of Society of Nature conservators (NATCON).

Dr. D.S. Malik, Ph.D., FNC, F ASEA, is senior lecturer in Department of Zoology and Environmental Science, Gurukula Kangri University, Hardwar (Uttaranchal), India. He is a well known researcher and academician in the field of 'Biodiversity Resource Conservation'. He has been published about 60 research papers, edited six books and also organized several national symposiums on different environmental themes. Dr. Malik is executive editor of 'Journal Aquacult'. Dr. Malik has been awarded Young Scientist Awards of ASEA (India), Nature Conservators, Paryavaran Award 2000 and Nature Conservators Impetus Gold Medal 2003 and associated with different National and International societies.

INTEGRATED FISH FARMING

By
Dr. Nityanand Pandey
Dr. D.S. Malik

2016
Daya Publishing House®
A Division of
Astral International Pvt. Ltd.
New Delhi - 110 002

First Published, 2008
Reprinted, 2016

ISBN: 978-81-7035-560-1 (Hardbound)

Publisher's note:

Every possible effort has been made to ensure that the information contained in this book is accurate at the time of going to press, and the publisher and author cannot accept responsibility for any errors or omissions, however caused. No responsibility for loss or damage occasioned to any person acting, or refraining from action, as a result of the material in this publication can be accepted by the editor, the publisher or the author. The Publisher is not associated with any product or vendor mentioned in the book. The contents of this work are intended to further general scientific research, understanding and discussion only. Readers should consult with a specialist where appropriate.

Every effort has been made to trace the owners of copyright material used in this book, if any. The author and the publisher will be grateful for any omission brought to their notice for acknowledgement in the future editions of the book.

Published by : **Daya Publishing House®**
A Division of
Astral International Pvt. Ltd.
– ISO 9001:2008 Certified Company –
4760-61/23, Ansari Road, Darya Ganj
New Delhi-110 002
Ph. 011-43549197, 23278134
E-mail: info@astralint.com
Website: www.astralint.com

Digitally Printed at Replika Press Pvt. Ltd.

Acknowledgements

It is my privilege to express my deep sense of gratitude, reverence, honour and sincere thanks to Dr. Davendra Singh Malik, Sr. Lecturer Department of Zoology and Environmental Science and my supervisor for his keen interest, expert guidance, constant encouragement, constructive criticism and invaluable help during the course of this study.

I express my heartful regards and profound gratitude to Prof. A.K. Chopra, Head, Department of Zoology and Environmental Science for his kind encouragement, valuable suggestions throughout the tenure of the research work.

I am deeply indebted to Prof. Virendra Arora, Principal, Science College and Prof. D.K. Maheshwari, Dean, Faculty of Life Science for his valuable guidance and inspirations during my studies.

I express my heartful thanks to Prof. B.D. Joshi, Dr. D. Bhatt, Dr. D.R. Khanna, Dr. P.C. Joshi and Dr. R.K. Negi for their moral support, unending encouragement and affections.

I am really grateful to Dr. Kapur Singh, chief training organizer, Krishi Vigyan Kendra, Rampura-Rewari to provide the facilities and support to carry out this research work.

With immense pleasure, I acknowledge the affectionate behaviour and timely help of my colleagues and friends namely

Ashutosh Mishra, Vipul Gupta, Vijay Kumar Singh, Dr. Pramod Kumar, Abhishek Shrivastava, Shiv Ji Singh and Sanjeev Kumar.

I am very obliged to my parents for their affection, love, sacrifices and blessings during the course of my study. I am highly indebted to my brother for their encouragement and affection.

My due appreciation to my wife Mrs. Renu Pandey, who by virtue of incessant inspiration, attachment and fore bearance has stimulated me to complete the work in the present form. My heartiest love and thank goes to my children Nityanshi and Aabhas who relieved me of tiredness by their sweet and innocent antics during the course of present study.

A word of appreciation goes to Mr. Sanjeev Kumar for putting his best efforts for typing the manuscript sincerely, neatly and timely.

I take this unique opportunity to acknowledge and express my gratitude to all those who have helped me in various ways during the present study.

Nityanand Pandey

Preface

Fish may be cheapest animal product, if waste replace the need of expensive supplementary fish feed. The potential to recycling organic waste into fish ponds has been demonstrated by the leading country of aquaculture production in the world, China where animal excreta is the main fertilizer in the fish culture. Integrated fish farming is generally considered particularly relevant to benefit the rural fisherman community. The integration of fish farming with livestock and crop farming offers greater efficiency in resource utilization, reduces risk by diversifying crops and provided additional income and employment opportunities to the fishermen family. In Asia, fish farming has been a part time activity of farmers, who developed it as an efficient means of utilizing farm resources to the maximum capacity. The highest productions obtained so far in integrated fish farming are with pigs, ducks and chicken, a very widespread technique in Asia. But, integrated livestock fish farming systems is more complex with a need for more skill and knowledge and better management. Integrated farming is a multicommodity farming system and failure of one system could adversely affect the other. Hence, there is a need to optimize the integrated livestock fish farming to get maximum production of animal protein per unit area at low cost with efficient management practices.

This book is a result of on field experimentation on *Azolla*-Pig-Fish farming and I hope that this book would be useful to the concerned and help in further improvement in management practices of integrated pig-fish farming.

N.N. Pandey

Contents

List of Abbreviations

BOD	:	Biological oxygen demand
DO	:	Dissolved oxygen
TDS	:	Total dissolved solids
per cent	:	Percentage
g	:	Gram
ppm	:	Part per million
ml	:	Milliliter
Ec	:	Electric conductivity
Mg/l	:	Milligram per liter
Fig.	:	Figure
°C	:	Degree centigrade
kg	:	Kilogram
kg/ha	:	Kilogram per hectare
N_2	:	Nitrogen
ppt	:	Part per thousand
K	:	Condition factor
C	:	Carbon

GPP	:	Gross Primary Production
NPP	:	Net Primary Production
CR	:	Community Respiration
log	:	Logarithmic
nos.	:	Numbers
cm	:	Centimeter
ds/m	:	Deci-siman per meter
@	:	At the rate

List of Figures

List of Plates

List of Tables

Chapter 1

Introduction

The world is facing a variety of interrelated problems such as food production, energy use, pollution, unemployment and urbanization. The current dilemma occurs because our strategies for development contravene the basic philosophy of ecology. The world ecosystem is becoming increasingly more unstable through an ever growing dependence on finite resources of fossil fuel. Food production system in developed countries, both terrestrial agriculture and fisheries are energy intensive and are thus not entirely suitable for developing countries. An alternative strategy to an increasing reliance on fossil for agriculture development should depend on locally available sources of renewable energy and resources. A significant effect could be made through recycling organic wastes into fish. Organic waste can be recycled in ecologically sound pond or lagoon systems fueled by solar energy. The reuse of wastes would safeguard the environment from pollution and lead to a reduction in diseases of insanitation. Recycling of animal wastes is the cheapest way of protein production by reducing cost of production.

Fish farming in Europe, Israel and USA is energy intensive, since more farmers use commercially prepared feed formulations which results in cost to feed often exceeding 50 per cent of the total operating costs (Schroeder, 1977).

In recycling of waste through fish culture, fish is considered to have a great potential to change raw protein into high energic edible protein. Fish may be the cheapest animal product, if waste replace the need of expensive supplementary feed. The potential to recycling animal wastes into fish pond has been demonstrated by the Chinese, where animal excreta is the main fertilizer in the fish culture. All farms undertake animal husbandry to supply fertilizer for fish farming (FAO, 1977). The rationale behind raising fish on organic manure is that about 72.98 per cent of the nitrogen, 62.87 per cent of phosphorus and 82.92 per cent of potassium in the feed rations fed to animals are recovered in their excreta. In most of the world, cereals and legumes account for most of the proteins and calories consumed, but fish produced cheaply by recycling wastes would provide a low energy way to produce an animal product and diversify the diet. It is now widely appreciated that the so called 'protein gap' is a myth and that persistent and widespread malnutrition in developing countries is due to a lack of sufficient food rather than insufficient good quality protein (Sukhatme, 1975).

In developing countries inland aquaculture is often integrated with other agricultural activities. The integration of aquaculture with livestock and crop farming offers greater efficiency in resource utilization, reduces risk by diversifying crops and provides additional food, income and employment. Thus, the three major side effects of population growth which are hunger, poverty and unemployment are solved through integrated fish farming. Integrated farming may be defined as an innovation, in which two or more commodities are framed together on a common infra-structural base with the objective to optimizing use of available resources. The basic principles involved in this system are the utilization of the synergetic effect of the interrelated farm activities and the conservation including the full utilization of farm wastes (Pillay, 1990). It is based on the concept that "there is no waste and waste is only a misplaced resources, which can become a valuable material for another product". There are several advantages of integrated farming system *viz.* increased productivity, greater income, improved cash flow, fuller employment, a better diet for the farmer's family and the less biological and economic risk. On the other hand, there is the disadvantage that the integrated system is more complex, with a need for more knowledge and better management, since the failure of one system could adversely affect the other. Hence, there is a need to optimize the integrated livestock fish farming to get maximum production of food per unit area with efficient management practices.

Integrated fish farming is generally considered particularly relevant to benefit the rural poor. In Asia, fish farming has been a part time activity of peasant farmers, who developed it as an efficient means of utilizing farm resources to the maximum capacity. The highest productions obtained so far in integrated fish farming are with pigs, ducks and chicken, a very widespread technique in Asia (Edwards *et al.,* 1986 and Edwards, 1983). In some countries, fish farmers also integrate geese, rabbits, goats, sheep, cattle and water buffalo with fish culture, on a smaller.

The daily excretion of organic waste by the animals varies depending on the age and the diet of the animal. The percent recoverable from the enormous wastes available also varies very much in each production system. In integrated livestock farming system, it is necessary to enclose the livestock so that the waste from feeding and the excreta (faeces and urine) can be collected adjacent to the fish pond to minimizes the transportation costs or built over the pond. The latter practice has the additional advantage of better utilizing the space where land availability is short.

Pig is one of the most prolific and efficient feed converting animal. In rural India, it is a traditional occupation of weaker section of society and an important source of income. Being a profitable business after adoption of improved technology, people of forward and middle class are also taking part in the pig production, which has given a boost to this enterprise in India. Raw pig effluent is offensive in terms of it's odour and also as a potential pollutant. Disposal of piggry wastes is a problem of most of the piggery farms. Pig waste is also good manure for agriculture and aquaculture practices. The application of the pig dung in fish ponds provides a nutrient base for dense bloom of phytoplankton and nanoplankton, which, in tune, form a base for intense zooplankton development. There is high production of detritus at the manured pond bottom, which provides substrate for micro-organisms and benthic fauna. On this basis, the more appropriate species of fish for a pig waste fed pond are therefore, those species which are able to filter and feed on plankton. The most common filter feeders are carps and mixed feeder on algae and detritus are certain species of Tilapia.

A very large proportion of the cost of pig production is invested for feeding. On an average 5 kg concentrate mixture is required to feed an adult pig daily. This cost can be minimized to provide feedstuff produced on the farm it self. Several aquatic plants like azolla are grown in the associated water near the farm. These aquatic plants as well as other foliage of several terrestrial plants, such as vegetables, corn, and rice can be utilized as pig feed.

Azolla is an aquatic fern with wide distribution all over the world having the capacity of assimilating atmospheric nitrogen through *Anabaena azolla*, a symbiotic blue green algae, present in the cavities of dorsal lobe of leaf. Azolla is also a bio-fertilizer for fish pond to enhance productivity. It also may be supplemented in the diet of carps. Pig dung may be recycled for azolla production. This floating aquatic weed may be used for aerobic digesion of piggery effluent to reduce the pollutant load.

Though several workers have determined the productivity of pig manured fish. But aerobic digestion of pig effluent before discharging in fish pond, its effect on over all productivity is quite meagre. No attempt has been done to integrate azolla in pig-cum-fish farming system as manure, feed and aerobic digester in Indian conditions. Hence, the aim of the present study was to evaluate an integrated farming system comprising livestock, azolla and fish. Pigs were raised adjacent to ponds. Their feed was supplemented by nitrogen rich aquatic fern azolla (*Azolla pinnata*) grown in the stabilization pond. Azolla treated semi-digested manure from stabilization pond was used to fertilize the fish pond. Fish pond was managed on zero supplementary feeding cost. This farming system is expected to generate a higher income per unit area than existing systems.

Chapter 2

Historical Review

Carps are widely cultivated fish species in Asia, both under mono and polyculture systems. In polyculture system several fish species are reared together in same pond that feed on different natural organisms to utilize the fish production potential of a pond. Polyculture, which is the traditional method of fish culture in Asia (Lin, 1982) started in China during the Tang dynasty (AD 618-907) with the joint culture of Chinese carps (Chang, 1987). Polyculture has expanded with the introduction of Chinese carps to many countries to obtained increased fish production (Milstein, 1990). In India, traditional polyculture of the Indian major carps (catla, rohu, and mrigal) has been changed by a system in which these indigenous species are cultured together with Chinese carps in appropriate proportions (Jhingran, 1986). This system is commonly referred as composite fish culture.

Success in composite fish culture depends largely on sustained production of fish food organisms. Organic and inorganic fertilizers are added to the fish ponds at periodic intervals to maintain the optimum level of essential nutrients.

Fish is the cheapest animal products when grown on wastes (Edwards, 1980). Fish grow rapidly in tropical water and if waste replace the need for expensive supplementary feed, the cost of production is minimal. The potential for recycling animal waste have been demonstrated by the Chinese who use animal manure as the main fertilizer in fish culture.

The prospects of fish culture in organic waste like cow-dung, poultry, duck, pig excreta and sewage have been suggested by Alikunhi, 1957; Lakshmana *et al.*, 1971; Grower *et al.*, 1976; Sen *et al.*, 1978 and Barash and Schroeder, 1983.

Hickling, 1962; Cruz and Shehadeh, 1980; Delmendo, 1980; Sin, 1980; Hopkins and Cruz, 1982; Jhingran and Sharma, 1986; Sharma and Olah, 1986, NACA, 1989; Pekar and Olah, 1991; Esteky *et al.*,1995; Zoccarato *et al.*, 1995 and Sharma *et al.*, 1998 have described culture technique and fish production in various crop-livestock integrated farming systems in countries like Hungary, Philippines, Malaysia, Germany, China, Taiwan, Madagascar and India.

Many authors have emphasized the importance of fish livestock integration in recycling of waste products, income generation and diversification of products (Woyanarovich, 1979; Yadav *et al.*, 1986; Little and Muir, 1987; Sharma and Das, 1988; Radhey Shyam, 1995; Kaunhog, 1996; Sharma *et al.*, 1998).

Swingle (1957) reported better growth of fish with higher rate of application of manure. Chen and Li (1980) described the Chinese AAA system (Aquaculture-Animal husbandry-Agriculture). This system recycles organic waste such as weeds and crop byproducts from field, livestock and natural food production from photosynthesis within the farm itself. Delmendo (1980) presented the review on integrated livestock-fowl-fish farming system. He also described the integrated farming system, being practiced in various countries of Asia, where animal house are erected right over the fish ponds. Pathak (1981) presented the data regarding energetics of an integrated farming system. They further reported that the production of autotrophic and heterotrophic in an aquaculture system help in trapping the solar energy. Barash and Schoeder (1983) tested the possibilities of replacing part of all of fish feed pellets with fermented cow manure. Sharma *et al.* (1985) presented the package of practices for fish livestock farming. Mishra *et al.* (1986) showed the possibilities of obtaining 52 per cent higher production of Indian major carps through the application of water hyacinth manure. He also indicated 70 per cent increase in production by combined application of hyacinth manure, biogas slurry and cattle shed washing. Patra and

Ray (1988) investigated the influence of a combination of pigeon dropping, goat dung and cow dung on growth and production of Indian major carps and suggested that cow dung, goat dung, pigeon dropping significantly increase the growth and production of Indian major carps. Singh and Das (1993) developed the model of integrated system or triple A (Aquaculture-Agriculture-Animal husbandry) system for marginal farmers. This system was based on the recycling of wastes. He calculated that a farmer having around 3 acres of land can earn Rs. 35,000-40,000 annually and that would be sufficient for him to run a small family. Fang *et al.* (1994) observed the effect of animal manure protein (Chicken, duck, pig, cow) on fish yield and reported that the conversion efficiency of manure protein into fish protein was about 40 per cent on a dry weight basis in the fish pond. Kestemont (1995) has made a critical review of various agro-aquaculture integrated system *viz.* direct or indirect and parallel or sequentional.

Different management aspects of organic fertilization in fish ponds have been reviewed by Edwards, 1980; Olah *et al.*, 1986; Wohlfarth and Schroeder, 1987 and Varadi, 1990.

The ranking order worked out by Ray and David (1966) and Govind *et al.* (1978) of different animal wastes potency was poultry > cowdung > goad > sheep > piggery. However, Kapur (1981); Kapur and Lal (1986 a,b) and Yadav (1987) found poultry > piggery > goat > sheep > cowdung. Prinsloo Schoonbee (1987a) noted the comparative effectiveness as duck manure > pig manure > raw chicken manure > cattle manure > sheep manure.

The FAO Technical conference on Aquaculture held in 1976 in Kyoto, Japan, devoted one of his sessions to integrated farming. The resolution of this fundamental global meeting, which has come to know as the "Kyoto declaration on Aquaculture", pointed out that "...... aquaculture can, in many circumstances, be combined with agriculture and animal husbandry with mutual advantage and contribute substantially to integrated rural development". The International center for Living Aquatic resource (ICLARM) organized an international conference on integrated agriculture, aquaculture farming systems in 1979 and promotes these even outside of Asia. The center on Integrated Rural development for Asia and the Pacific (CIRDAP) initiated studies in 1986 on integrated farming and organized a regional workshop as a follow up in 1988. The network of Aquaculture centers in Asia (NACA) assigned one of its four regional centers for integrated fish farming and published a text book on this subject in 1989.

Pig fish integration is one of the excellent way of recycling of all the organic waste efficiently in fish pond as a source of nutrients. Nutrients requirement of fish pond which depends mainly on the nutrients status of pond soil and fish density there in, can be fulfilled by supplying needed quantity of excreta by regulating the number of pigs stocked with pond.

Experiment on integration of fish culture with piggery was initiated at CIFRI, Barrackpore during 1978 (Sharma *et al.*, 1979). Integrated fish farming by recycling of pig manure in fish pond have been reported by Sharma *et al.*, 1979; Cruz and Shehadeh 1980; Woynarovich, 1980; Sharma *et al.*, 1985; Sharma and Olaha, 1986; Sharma and Das, 1988; Gavina, 1994 and Borah *et al.*, 1998 in India and abroad.

Ashwathanarayana (1979) has compared the effectiveness poultry manure, sheep and goat manure and pig dung on carp production and found poultry and sheep manure as equally effective. Woyanarovich (1979) Obtained a yield of 15-18 t/ha/year from the water treated with duckery and pig manure. Kapur (1984a) reported the 20 per cent increase in fish yield with poultry-piggery waste combination in a ratio of 1:1 as compared to poultry waste alone. Sharma and Olah (1986) achieved 40 kg/ha/day fish production, stocking 18000 fish/ha under polyculture with aquatic plants and pig manure as the only organic inputs to the pond. Olah *et al.* (1986) studied 54 ponds and found high range at stable primary production (5.3 to 6.5 g cm^{-2}/day) in the pond, which were fertilized with oxidized liquid pig manure. Prinsloo and Schoonbee (1986) investigated the use of pig manure as nutrient to stimulate pond productivity in polyculture system of European common carp, big head carp, black carp, grass carp, silver carp and tilapia. Fish production was approximately five times higher in ponds receiving pig manure with formulated feed compare to the pond receiving pig manure alone. The pond water effluent was also used for vegetable production. Han and Ding (1986) studied the use of biogas fermented pig excreta as a manure source in fish ponds. Application of fermented manure resulted in the proliferation of phytoplankton. The fish yield was found to be 7.1-25.6 per cent higher than that by using fresh pig manure. Yang-Yijing *et al.* (1987) concluded that fish in fresh pig manure ponds not only could utilized bacteria, detritus and plankton but could also directly feed on pig manure. Zhu-yun *et al.* (1987) studied the frequency of fermented pig manure application and chemical fertilizer in pond stocked with carp species. Daily manured pond showed higher fish yield than those manured at 5 days and 7 days intervals. Fang *et al.* (1988) concluded from their study that 1 kg. of filter feeding and omnivorous fishes can be produced by using 27 kg of pig manure. Dutta and Goswami (1988) suggested that weekly manuring of mix cow and pig shed wastes at the rate of 10 tones/ha yielded three times greater weight of fish with out use of supplementary feed over control ponds. Billard *et al.* (1989) studied the use aquatic food chains for bioconversion of pig waste in Italy. Pig manure was anaerobically digested and used for sequential cultures of green microalgae, cladocerans and macrophytes. The system was found economic and significantly productive with effective recycling of pig manure. Pillay (1990) reported 2-18 t of fish and 4-7 t of pigs (live weight) per ha/year with stocking of 60,000 fingerlings of different species raised together with about 45-75 pigs/ha. Zhu-yun *et al.* (1990) studied the response of polyculture of silver carp, big head carp, common carp and crucian carp to varying amounts and frequencies of fermented pig manure. Net fish production was found to be directly proportional to the rate of microbial decomposition of cellulose at the sediment surface. Sharma and Saini (1992) studied the effect of different rates of pig manure on zooplankton production and concluded that very high or low dose of pig manure do not give better production. Pig manure @400 mg/l gave high production of zooplankton. However, water quality parameters were not significantly affected by pig manure fertilization when applied at optimum rate. Jaing *et al.* (1993) described the integration of fish farming, crops and animal husbandry in Dong ping lake region and suggested that by-products of grains could be used to feed pigs, pig manure could be use to feed fish and fish manure could be used as fertilizers for crops. Gavina (1994) indicated use of pig manure to fuel the natural food chain in the pond to enhance tilapia and azolla production.

The organic production in aquatic systems depends directly on the water quality, which intern is influenced by the bottom soil (Singh, 1979). According to Banerjea (1967), the nutrient status of water and soil play the most important role in governing the production of planktonic organisms. Thus, seasonal fluctuations of various physico-chemical factors of water and soil have an important role in distribution, periodicity as well as qualitative and quantitative composition of biota.

The soil of fish ponds is greatly altered due to continuous addition of the organic matter in the form of inputs such as animal excreta, fish feeds etc. (Chattejee and Saha, 2000). pH of the soil influences transformation of soluble phosphate and controls the absorption and release of ions of essential nutrients at the soil water interface. Moderate alkaline nature of the pond bottom soil provides conducive conditions for higher fish production (Banerjea, 1967; Jhingran, 1988 and 1991). Jhingran (1989) observed maximum productivity in ponds having organic carbon level of pond soil between 1.5 and 2.5 per cent. Humus and its alkali soluble fractions increase in soil of undrainable fish pond with the increase in organic carbon content (Chatterjee and Saha, 1996). Saha (1985) found seasonal variations in nitrogen concentration of fish pond soil due to addition of the nitrogenous compounds. Banerjea (1967) has attempted to correlate fish production with the nitrogen in soil of fresh water fish ponds. Jhingran (1988 and 1989) classified the productivity status of fish ponds on the basis of phosphorus content of pond soil. Chang and Jackson (1957) have described the factors influencing the content of various forms of phosphorus in the soil. The deposition of higher concentration of phosphorus in pond soil due to continuous addition of fertilizers reduces the ability of phosphorus removal of fertilizers from water (Eren *et al.*, 1977). Sandy non-absorptive soils have poor concentration of potassium (Jhingran, 1991). Concentration of calcium and phosphorus in soil is greatly influenced by the pH; calcium and magnesium ions remain in bound form with phosphorus at low pH. Banerjea (1967) had attempted to establish correlation between calcium in soil with pond productivity. Press *et al.* (1996) reported increase in soil phosphorus, potassium, calcium and magnesium by a factor of 2 or 3 by application of organic inputs. The carbon-nitrogen (C/N) ratio is an important factor in influencing the microbial decomposition of soil organic matter. Mohanty *et al.* (1994) observed decline in mineralization of organic nitrogen with the increase in C/N ratio.

The physico-chemical properties of pond soil have great bearing on the hydrobiological characteristics of pond water. Jone (1990) said that the temperature, pH, dissolved oxygen, turbidity, plankton, total alkalinity, total hardness and ammonia are the main parameters of water quality management.

Philipose (1960) pointed that best water for support of fish and aquatic life are those which have pH values ranging between 7-8. Banerjee *et al.* (1979) reported wide fluctuation in pH values in cow dung + poultry + wolffia as compare to cow dung alone treated water. Further, they suggested that the fluctuations in pH were mainly due to the higher metabolic rates of organic wastes and chemical transformation, which could super side the ecological buffering action of the water and hence they are ore productive because of production and assimilation of CO_2 at higher rate. Alabaster and Lloyd (1980) suggested the safe level of pH for fish ranging 5-9 and 6-5 to 8.5 for maximum productivity. Dinesh *et al.* (1986) found higher pH values in manured ponds and observed inverse relationship carbonate and bicarbonate alkalinity. Das (1987); Guiterez and Guevara (1988) reported that the organic matter increases the pH value of fish pond water. Galimoni and Negokaka (1989) reported the acidification of pH in pig manured ponds and basification of pH in poultry manured pond. Sharma and Saini (1992) reported a gradual decrease in pH value just after fertilization with piggery waste and subsequent increase after 15 days of application. Swaranalatha and Rao (1997) observed the interrelationship of certain physico-chemical factors and reported inverse relationship between pH and bicarbonates, dissolved oxygen and organic matter and direct relationship between nitrate and dissolved oxygen. The significant correlation between pH and productivity waste treated water have been calculated by Olah *et al.* (1986); Yadav (1987); Yadav and Kapil (1990).

Temperature is one of the critical environmental factors in the aquatic systems. It directly affects the various activities and metabolic processes of organisms (Hutchinson, 1957). The degree of annual variation in temperature of water body has a great bearing upon its productivity. The rate of decomposition of organic matter is highly influenced by the temperature regime of a water body (Karause, 1962). Philipose (1960) suggested that a high value (>30°C) and very low temperature affect production adversely. Hutchinson (1967) observed chemical as well as biological factors with changes in temperature and reported that water temperature significantly affects the biological productivity of ponds by controlling chemical and metabolic activity of organisms. Karause (1962) reported that the changes in temperature of a water body can influence the speed and course of decomposition of organic matter. Boyd and Pillai (1984) reported that the water temperature upto 32°C in tropical fish ponds dose not seem to adversely affect the production of warm water fish. Alam *et al.* (1990) revealed that higher temperature (28-30°C) showed better production of primary production due to nutrient availability hence good fish growth. Khanna (1993) reported the reduced metabolic activities at low temperature.

Dissolved oxygen is the most critical factor for the survival of aquatic life. Dissolved oxygen regulates the metabolic processes of aquatic organisms including fish and is often considered as an indicator of environmental conditions affecting survival and growth of organism. Oxygen in water is dissolved either through the diffusion from the atmosphere and/or evolved in the pond itself through the process of photosynthesis by primary producers mainly, phytoplankton in fish ponds. Dissolved oxygen concentration of 6-9 mg/l in pond water is considered optimum for the proper growth and survival of fish (Boyd, 1982). Low concentration of dissolved oxygen in water causes suffocation in fishes while its super saturation may result into the gas bubble disease leading to the mass mortality of fishes in both the cases. Rao and Raju (1989) and Nath *et al.* (1994) have recorded the dissolved oxygen values ranging between 2.6-9.4 mg/l in semi-intensive carp culture ponds. According to Odum (1971), a daily net production of 8 g O_2 m^{-2} indicates a balanced nature of an ecosystem.

In pond ecosystem, addition of high nutrient load (organic matter) can harm fish health by depleting dissolved oxygen in early morning hours (Gill, 1974; Sehroeder, 1974; Yadav, 1979; Boyd, 1979; Das, 1987, Fang *et. al.* 1988 and Guiterez and Guevara, 1988).

Philipose (1960) showed that DO range of 4–10 ppm was favorable to carp culture. Doudroff and Dean (1967) have suggested the lowest limit of 5 ppm dissolved oxygen in tropical and sub tropical ponds for maintaining the healthy crop of organisms. Similar results were also reported by Kapur (1984a) and Yadav (1987). Banerjee *et al.* (1979) Compared the efficiency of poultry excreta and cowdung in terms of their ratio of mineralization at different temperature and observed that the higher (>10 kg/ha) dose of animal waste may adversely affect the diurnal oxygen budget. Yadav (1988) reported the increased toxicity level of effluent, mainly due to the reduction of dissolved oxygen levels in pond water. Sharma and Saini (1992) reported the decrease concentration of dissolved oxygen with increased water temperature.

Oxygen balance of water is affected by biological oxygen demand (BOD) when the environment is rich in organic matter. BOD is a measure of the amount of oxygen required by microorganisms to decompose the organic matter in water body (Boyd, 1978). Schroeder (1975) reported the significant correlation between temperature and BOD load of waste treated water and also noticed that BOD increased more than two folds with increase in temperature from 10°C to 20°C. However the less *i.e.* below two folds increase in BOD was noticed when the water temperature increased from 20 to 30°C. similar results have also been obtained by Vashisht and Shekhar, 1983; Kapur, 1984; Lal, 1986 and Lal and Kapur, 1987. Yadav (1987) observed the increased dissolved oxygen and decreased BOD from 27.5 and 50°C temperature. Banerjee *et al.* (1979) showed that BOD value is a primary indicator of organic pollution in waste body, therefore discharge of excess of organic waste exert load of BOD. Mora (1982) has classified water as fresh, medium and polluted on the basis of BOD. Vashisht and Shekhar (1983) reported the decreased BOD level and increased DO concentration in waste treated water due to the microbial decomposition of waste.

The processes involving bacterial decomposition of organic matter and respiration by the biota are the chief sources of CO_2 levels in fish ponds. Carbon dioxide exists in the ponds in two states, namely bicarbonate and carbonate, serving as a more or less purely chemical sources needed to buffer the environment against rapid shifts in acidity-alkalinity status. Secondly, the levels of CO_2 regulate the biological processes in aquatic communities. The third and most important contribution of CO_2 is that it serves as an important source of organic carbon (Reid, 1961) for biota. Ellis (1937) stated that the water supporting good fish population usually contains less than 5 ppm of free CO_2. Parks *et al.* (1975) stated that the free CO_2 levels fluctuate between 0 ppm during after noon and around 5-10 ppm during dawn with out any apparent ill effect on fish. Boyd (1978) pointed out that, the most detrimental influence of free CO_2 in fish culture may result when CO_2 concentration increases during critically low DO levels. Srisumantach (1982) pointed out that low algal production resulted in accumulation of CO_2 in water. Boyd and Pillai (1984) stated that uptake or release of CO_2 by organism may change the proportion of carbonate and bicarbonate ions in water.

Alkalinity is the capacity of pond water to neutralize the acids shifting the pH value towards the alkaline side. Several workers have reported that the main contributors to alkalinity in natural waters are bicarbonate, carbonate and hydroxide. Srisumantach (1982) reported the increasing trend of alkalinity in manured fish pond. Tripathi (1982) noticed the alkalinity range of 100-120 ppm for better fish growth. Dinesh *et al.* (1986) showed increase relationship between carbonate and bicarbonate alkalinity. Swarnalatha and Rao (1987) reported inverse relationship between pH and bicarbonates, dissolved oxygen and organic matter. Singh and Sharma (1998) showed a reverse trend of carbonate alkalinity against free Co_2 concentration in animal wastes treated ponds.

Conductivity of water depends on level of dissolved salts. A higher value of conductivity in organic waste treated pond reported by Gill, 1984. Ghosh *et al.* (1973) observed 10-12.5 ppt salinity level to be critical for survival of Indian major carps. Chattopadhyay (1998) described the 0.5-1.0 ppt salt content in fresh water and pointed out that most of the fishes of fresh water grow well under water salinity upto 3.0 ppt.

An approximation of EDS value of water can be made by multiplying the respective EC value with a factor, Boyd (1978) suggested that this factor may be ranging between 0.55 and 0.90 depending on the composition of the dissolved solids.

Ammonia is an important nitrogenous end product of heterotrophic break down of organic matter. Certain amount of ammonia also comes through the excretory waste products of aquatic animals. The most important source of ammonia is the ammonification of organic matter in water (Armstrong, 1979). The higher concentrations of ammonia are harmful to fish and other biota. The increase in water pH decreases its toxicity due to conversion of ammonia into ammonium ions which are much less toxic than the gaseous form (Downing and Merkens, 1955). Ammonia becomes more toxic when the levels of dissolved oxygen decrease (Merkens and Downing, 1957). Boyd and Pillai (1984) recommended the maximum acceptable level of ammonia in fish ponds as 1.0 ppm. Rana *et al.* (1996) found positive correlation between ammonia and fish growth. Akpan and Okafor (1997) estimated the increasing value of ammonia as 2.0-2.4 ppm in poultry manured pond.

Nitrite and nitrate are the oxidized forms of ammonia nitrogen and are important plant nutrients. Padmavathi *et al.* (1997) reported that generally nitrate occurred in very high concentrations in pond water, while ammonia ranked next to it. Lyubumova (1974) reported level of No_3 as above 3 ppm to ensure good production of autotrops. Shankar (1978) and Hegde (1981) revealed that the concentration of nitrates in livestock treated waters, increases just after a few days of treatment. Stewart *et al.* (1992) studied and suggested the over dose of manure in fish ponds could result in undesirable salts accumulation and leaching of nitrates to ground water. Anuta (1995) observed the increasing concentration of nitrate, nitrogen, ammonia nitrogen in the water with increasing amount of manure application.

Nutrient budget provides vital information on the utilization pattern and efficiency of the inputs supplied to the fish ponds. A number of investigations have been carried out to calculate the budget of important nutrients such as nitrogen, phosphorus and carbon (Parnas and Lahav, 1972; Worsham, 1975; Avnimelech and Lacher, 1979; Boyd, 1985; Krom *et al.*, 1985; Schroeder, 1987; Daniels and Boyd, 1989; Krom and Neori, 1989; Briggs and Funge-Smith, 1994; Martin *et al.*, 1998). The major source of nutrients in ponds is manure/fertilizers and fish feed. Fish feed accounted up to 92 per cent of the gross nitrogen input in commercial scale aquaculture (Briggs and Funge-Smith, 1994; Marco *et al.*, 1994).

Among the various inorganic nutrients, phosphorus has a unique role in the development of autotrophic aquatic communities (Boyd, 1989). According to Wetzel (1975) phosphorus limits the biological productivity if its concentration is low. However, heavy precipitation of phosphorus in system could mask the phosphorus input through left out feed, excreta and fertilizers (Padmavathi *et al.*, 1997).

McCombie (1953) reported direct relationship between phosphorus content and plankton volume. Banerjea (1967) studied concentrations of phosphorus in numbers of fish ponds and suggested that >2.0 ppm of P_2O_5 in pond water may be considered to be optimum while 0.05–2.0 ppm may be considered to be moderate and <0.05 ppm to be poor. Quasim *et al.* (1969) concluded a positive correlation between phosphorus and primary productivity in manure fed fish pond. Golachowaska (1971) studied the pathways of phosphorus in water and found that the production ability is determined not only by the absolute level of phosphorus but also by the forms in which it occurred.

Potassium is least important plant nutrient and its requirement is generally fulfilled by the pond itself. Potassium concentration in natural waters usually ranges between 0.5 and 10 mg/l (Moyle, 1946; Arch and Boyd, 1980).

Philipose (1959) stated that the production of autotrophic organism is the best index for the estimation of nutritive condition and biological production in animal waste treated ponds. Studies on plankton. Benthos community in fish pond in relation to productivity have been done by Patnaik *et al.,* 1988; Dhawan and Toor 1989; Haque *et al.,* 1990; Singh and Kumari 1999; Mukerjee and Pankajakshi 1995; Singh and Sharma 1998; Chaudhari and Singh 1999.

The minute free floating organisms, whose power of locomotion is so feeble that they remain at the mercy of water currents are known as the plankton. Welch (1952) has defined plankton as the heterogenous assemblage of minute organisms occurring in water columns. The planktonic organisms of plant and animal origin are called phytoplankton and zooplankton, respectively. The fish yield of almost all aquatic ecosystems is directly or indirectly dependent on plankton abundance. The information on distribution, composition and succession of plankton provides valuable information for determining the suitable species for stocking and understanding the level of utilization of the available food by fishes. The study of plankton is, thus, essential for efficient management of water bodies.

Primary production is the elaboration of organic matter from inorganic materials within a certain period of time by autotrophic organisms in the presence of radiant energy. Among the various fresh water plant communities, phytoplankton constitutes the major component of primary production. Primary productivity has been used as a potential index for describing the functional aspects of many diverse ecosystems of the world (Wetzel, 1966).

According to Ruttner (1953), gross primary production is the total amount of organic matter produced in a given time including that used simultaneously in metabolic processes. Net primary production is the assimilation surplus in a given period of time and can be quantified by subtracting of dissimilation from assimilation. The species richness, density, chlorophyll *a* concentration and primary production of phytoplankton are greatly affected by solar radiation (Wetzel, 1966 and Efford, 1967), temperature (Sreenivasan, 1976; Roa *et al.*, 1982), pH (Rao *et al.*, 1982), free CO_2 (Mukherjee and Pankajkshi, 1995) and nutrient status of water bodies (Wright, 1959; Efford, 1967; Khan and Sidique, 1971; Sharma, 1980; Pant *et al.*, 1981; Rao and Raju, 1989 and Ayyappan *et al.*, 1990). Increase in the concentration of essential nutrients enhances the phytoplankton productivity (Rao and Raju, 1989; Ayyappan *et al.*, 1990).

Eskinazi and Koening (1980) observed that phosphate is the most important limiting factor for the phytoplankton and its addition in large quantity may cause blooms of algae. Dhawan and Toor (1989) propounded that growth rate of phytoplankton, zooplankton and fish were significantly higher in ponds receiving poultry dropping alone and in combination with cowdung. Haque *et al.* (1990) reported bimodal pattern of seasonal variation in phytoplankton population and obtained inverse relationship between phytoplankton density and transparency, phosphate phosphorus and pH of water. On the other hand positive relationship were recorded between phytoplankton, dissolved oxygen and zooplankton. Singh and kumari (1994) observed that each group of phytoplankton showed different peak periods. The temperature, pH and dissolved oxygen found to be direct related with phytoplankton. Mukherjee and Pankajkshi (1995) observed that excess concentration of phosphates and nitrates caused dominance of microcystis. Magizharasan and Indra (1997) recorded seasonal fluctuations in phytoplankton population with peak population during summer. The productivity of zooplankton was higher in winter. Baruah *et at.* (1998) reported eighteen lighten species of phytoplankton belonging to chlarophyceae and mixophyceae in the fish culture ponds.

Zooplankton are the grazers of phytoplankton therefore, their population depends on the availability of phytoplankton (Sivakami *et al.*, 1995). The diversity of zooplankton is generally higher in temperate waters as compared to subtropical and tropical waters (Lewis, 1979; Sharma, 1980; Singh *et al.*, 1990 and Rawat, 1991).

Light, temperature, dissolved oxygen, nutrient and food are major factors affecting zooplankton abundance (MicIntire and Boyd, 1962; Wetzel, 1975; Sivakami *et al.*, 1995). A number of investigations have been carried out on the species composition and seasonal variation of zooplankton in fresh water ecosystem (Singh and Sharma, 1998).

Patil and Gouder (1985) reported two peaks of zooplankton production in fish ponds, one during winter and other in summer. Dutta *et al.* (1987) observed that zooplankton community was mainly dominated by rotifers followed by copepods and cladocerans in majority of water bodies. Bhattacharya *et al.* (1988) reported that phytoplankton were generally abundant during day time whereas zooplankton were more common during night hours in the surface water. Patnaik *et al.* (1988) noticed that the abundance peak of zooplankton were recorded during mansoon and winter. Ahmad and Singh (1991) recorded positive correlation between rotifer and copepods populations whereas cladocerans showed no such correlation. Pushpendra and Madhyastha (1994) recorded maximum number of zooplankton species in September, although maximum population density occurred during November, February and April. Decreased density of zooplankton is attributed to the acidic conditions of the pond. Jana *et al.* (1997) found that frequent application of low doses of manure resulted in significantly higher number and biomass of Daphnia than the monthly application of the manure. Shuter and Ing (1997) observed a simple mode of the seasonal production cycle for limnetic zooplankton in which weight specific rates of biomass accumulation were set by food resource availability and body size. Singh and Sharma (1998) obtained higher plankton number in poultry excreta and Pig dung manured ponds as compare to cowdung manured ponds. Brachionus, Keratella, Daphnia and Cyclops were dominant zooplankton species. They reported plankton volume in the range of 0-0.01 cc/l and zooplankton density ranging from 558 nos./l in waste manured pond. Chaudhari and Singh (1999) noticed that rotifers were dominant over other groups in all seasons. Pandey and Malik (2002) studied zooplankton production with application of pigdung biogas slurry and observed higher population in comparison to fresh pigdung.

Macrobenthic invertebrate fauna, such as larvae and pupae of chironomids, oligochaets and molluscs constitute the main food of benthophagus fish (Elango *et al.*, 1997). Mostly they are herbivorous/ detritivorous and constitute the second trophic level in the food chain. The study of benthic community forms an essential component for understanding the production of higher trophic levels in aquatic systems.

Pandey *et al.* (1983) observed that the molluscs insects, ostracods, annelids, nematodes and crustaceans are the main group of zoobenthos in Bakhira lake, Molluscs showed a positive correlation with temperature and negative correlation with insect density. Raghavan *et al.* (1986) noticed high population of chironomids in the benthic fauna in two village ponds. Veitenheimer *et al.* (1990) recorded higher population of chironomid larvae, molluscs, oligochaets and trichopteran larvae in manured pond. Oertli (1995) noticed that zoo benthos densities were largely dominated on all substrates and seasons by chironmids and tubificides. On all substrates, the highest monthly densities were observed in summer or beginning of autumn in a pond. Singh and Singh (1996) obtained highest number of zoobenthos during rainy season (August) and lowest in winter months (January).

Raw pig effluent is offensive in terms of its odour and also as a potential pollutant in water courses. The oxidation of its organic matter content in water courses uses up oxygen in the water and as a consequence, places at risk fish and other aquatic life. Use of treated effluent is batter safer than the raw pig effluent. Various methods are in vogue for the treatment of wastes containing organic and inorganic pollutants. These processes are based on the principles of precipitation, coagulation, chemical oxidation, sedimentation, filtration and adsorption.

Work on biological treatment of effluent has been done by Patrick 1953, Cairns, 1965; Sinha and Sinha, 1969; Shroff, 1977; John, 1984; Trivedi and Knourane, 1985 and Ali and Newton, 1988.

Macrophytes based system came to light during the mid 1970s from scientists working in Germany (Kickurth 1975, Seidel 1976, Seidel *et al.* (1978), Netherlands (De Jong, 1976, Kok, 1974) and USA (Dinges, 1976; Wolverton and Mc Donald, 1975a 1975b). Oron *et al.* (1984) stated that duckweed such as *Lamna gibba* has been utilized to treat municipal waste water in Israel and a very high BOD removal (97 per cent) has been reported. John (1986) pointed that even a poorly designed aquatic macrophyte based treatment system (AMS) can satisfactorily remove many waste water contaminants. Abbasi (1987) reported that in rural areas, barns are built in such a way that the manure from dairy animals falls directly into ponds to fertilize crops of aquatic plants, which has been harvested regularly and provided as food for animals. Abbasi *et al.* (1988) stated that aquatic macrophytes can also be very effectively used in thin film system for rapid and efficient contamination removal. Fallow field and Martin (1988) introduced an aerobic algal treatment system for piggery waste. English *et al.* (1989) reviewed pig effluent storage treatment and disposal. Yeoh (1993) enhance the final discharge quality of some agro-industrial waste water treated in lagoon systems by using water hyacinth. In the secondary treatment of small scale piggery effluent, the aquatic weed was found to be 17-112 per cent more efficient than algae in removing the BOD. Flessa (1994) reported that a portion of transported oxygen is used by aquatic macrophytes during root respiration, the rest enters in the water column and is utilized by facultatative and aerobic bacteria for the oxidation of organic carbon. This ability of aquatic plants to 'pumpin' oxygen can be strong enough to enable oxidation of even strongly reduced sediments in eutrophic water bodies. Weisner *et al.* (1994) stated that in AMS, BOD is utilized by bacteria as an energy source and for cell synthesis. Aerobic bacteria utilize oxygen as an electron acceptor in the break down of substrate carbon, where as facultative anaerobic bacteria utilize oxidized inorganic compound, such as nitrates and sulphates as electron acceptors.

Patra and Santra (1999) reported that bioremediation of paper mill effluent using water hyacinth is most significant. The per cent removal of BOD, COD, Phosphate nitrate and TSS at 21st day was about 36 per cent, 40 per cent 58 per cent and 75 per cent respectively.

Chauhan *et al.* (1997) studied propagating azolla-cum-fish culture with use of distillery effluents and obtained that culturing azolla initially helps in the biological filtration of effluent.

Azolla is a heterosporous free-floating aquatic fern that lives symbiotically with the nitrogen-fixing algae, *Anabaena azollae* (Moor 1969, Peters *et al.*, 1979) Moore (1969) assigned azolla to a monotypic family Azollaceae and also reported 0.222 mg nitrogen fixed/day / g dry weight by azolla, suggested optimum temperature for growth and nitrogen fixation of *A. pinnata* as 20-30°C and also noticed that total N_2 requirement of the frond is supplied by algae. Ashton and Walmsley (1975) stated that the algae remains closely associated with apical meristem and grows in unison with the plants and is never indirect contact with external environment. Espinas and Watanabe (1976) reported N_2 fixation by azolla as 0.4-0.85 mg/ day/ g. fresh weight. Becking (1979) described the two morphologically distinct forms of *A. pinnata i.e. A. pinnata var. Pinnata* and *A. pinnata var. imbricata.* Lumpkin and Plucknett (1980) described that *A. pinnata* comprises the subgenus Rhizosperma and distributed in the most part of Asia and tropical Africa and also reported phosphorus as critical nutrient in the growth and multiplication of azolla. Sounders and Flower (1992) devided *A. pinnata* in to three sub species *viz. A. pinnata pinnata, A. pinnata asiatica* and *A. pinnata africana,* of which only *A. pinnata asiatica* occurs in India. Lumpkin (1977) observed death of azolla at 40°C temperature and brown raddish colouration below 20°C. Kannaiyan (1979) reported wide salt tolerance capability of azolla and suggested optimum temperature for growth and multiplication of *A. pinnata* as 24-30°C. Tung and Watanabe (1983) reported the heat tolerant strains of *A. pinnata* and *A. microphylla* showed good growth at an average temperature of 33°C. Uheda *et al.* (1999) reported the highest temperature tolerance capability of *A. pinnata* among 6 species tested. Ashton (1974) observed the fastest growth of Azolla in the PH range 4.5-7.0 but Azolla can survive from PH 3.5-10. Liu (1979) noticed that the availability of organic-N to Crops from Azolla green manure depends upon its C : N ration. Azolla decomposes and supplies-N most rapidly if C : N ratio is about 10 per cent. Peter *et al.* (1979) noticed the N-content of Azolla as 3-5 per cent dry weight (19-31 per cent crude protein) during optimum conditions of growth. Watanabe *et al.* (1980) observed 1.9-2.5 kg fixation of N_2 by azolla with 1kg application of phosphorus. Ito and Watanabe (1983) reported that when, Azolla is grown in media containing combined nitrogen, the Anabaena symbiont still fixes atmospheric nitrogen and supplies it to the host plants, and that each nitrogen source has specific incorporation and translocation pathway. Peters and calvert (1984) observed that the buds located at the ends of the plants develop in primary branches and then form secondary branches. Primary branches separate from parent to form a new plant. Zheng *et al.* (1999) studied genetic diversity and classified cyanobacteria in different azolla species by the use of PCR.

Azolla as an atmospheric N_2 fixer, water purifier, poultry feed, crop feed, substitute of fish meal upto 25 per cent, substitute of soybean, bio fertilizer, producer of hydrogen gas and useful in mosquitoes control is well documented by Galston (1975), Gopal (1967), Kasmi and Trivedi (1978) Wang and Fuller, Mandal and Ghosh (1990), Ayyappan *et al.* (1991), El-sayed (1992) and Kant *et al.* (1996). The sole variety of azolla, *A. pinnata* has been cultivated for agricultural purpose in Vietnam (Tuan Thuyet 1979), China (Liuchung Chu 1979), Thailand (Swatdee and Sectanum 1979), Philippines (Watanabe *et al.*, 1977) Indonesia (Partohardjono *et al.*, 1983) and in India (Singh 1977).

Because azolla has a higher crude protein content (ranging from 19-31 per cent) than most green forage crops and aquatic macrophytes and a rather favorable essential amino acid (EAA) compostion for animal nutrition (rich in lysine), it has also attracted the attention of livestock, poultry and fish farmers (Alejar *et al.*, 1984; De Waha Baillonville *et al.*, 1984a; Chen and Huang, 1984; Antoine *et al.*, 1987; Liu, 1987a Zhang *et al.*, 1987, Micha *et al.*, 1988; Santiago *et al.*, 1988). Wang and Fuller (1989) reported better amino acid profile in azolla than soyabean. Lumpkin and Plucknett (1982) and Van Hove and Lopez. (1983) suggested *Azolla filiculoides* as a source of protein in the diet of mogastric animals.

Van Hove C. (1989) and Cagauna and Pullin (1994) suggested azolla in animal diet as a source of protein. Maricel Becerra *et al.* (1990) studied on *A. filiculoides* and suggested as partial replacement for traditional supplements in pig diet and also reported decreasing performance of grower pigs with azolla supplemented diet but strong tendency for grow faster in fishing phase with overall good growth. Ayyappan *et al.* (1991) suggested that Azolla also have potentialities as a nitrogen biofertilizers in carp production. Gavina (1994) reposted good growth performance of pigs, duck and fish as affected by different levels of azolla meal in their diet. Shiomi and Kitoh (1996) reported the highest tolerance of *Azolla pinnata* against ammonia and highest capability of nutrient removal in May, than the other varieties of azolla. Basudha and Vishwanath (1997) reported higher tolerance of fish fry with formulated azolla based diet than growth of fish reared in manured pond. Fasakin and Balogun (1998) showed the total replacement of soybean meal by heat- treated azolla meal in the diet of fingerlings of *C. gariepinus*. Van Hove C and Lejeune

(2002) suggested use of azolla in integrated farming system manly in rice- fish-azolla and pig-poultry-fish-azolla and also stated that azolla is one of the most nutritive aquatic plants, owing to its high crude protein and carotenoid contents and generally good amino acid profile and can be incorporated into the feed of fish, pig, poultry, rabbit and even humans.

The integrated farming systems are potentially important in raising the income level as well as the standard of living of small-scale farmers (Vincke, 1991).

Some of the economic aspects of integrated fish farming have described by several authors (Vincke, 1979; Delmendo, 1980; Sin, 1980; Tan and Khoo, 1980; Djajadiredja *et al.*, 1980; Shang and Cost-Pierce, 1983; Edwards *et al.*, 1986; Lovshin *et al.*, 1986).

Shang and Costa-Pierce (1983) stressed that most of the economic aspects concentrate on rudimentary budget analysis, which estimates casts of production and profit of operation. Such studies usually provide little sensitivity analysis in relation to variations of productions input cost and market price.

Borah *et al.*, 1998 studied the economic aspects of integrated Poultry-Pig-Fish farming and reported 40.5 per cent decrease in the total variable cost with 306.7 per cent higher net economic return in the integrated unit in comparison to not integrated unit.

Chapter 3

Physio-Geographical Features and Design of Experimental Units

Field experiment was carried out at fish production and pig breeding farm of Krishi Vigyan Kendra (KVK), Rampura, Rewari (Haryana). Most of the laboratory work was conducted at KVK Rewari and some samples were analyzed at GKU, Haridwar. The experiment was conducted for 12 months (November 2002 to October 2003).

Geographical Location of KVK, Rampura

Rampura is situated in district Rewari, bordering Mohindergarh district of Haryana in the west and Alwar district of Rajasthan in the south-east. Rewari is located in the south western part of Haryana. It lies between 27.5° and 28.5°N latitudes and between 76° and 76.5°E longitudes. The soil of this area is sandy loam and slight alkaline in nature. The under ground water of most of the area is brackish with electric conductivity ranges from 1000 to 9000 ds/m. The climate of the area is almost semi-arid. Krishi Vigyan Kendra is located at 4 kms from Rewari city, on the link road (Rewari-Daliakee). The integrated pig-fish unit is located in main farm of KVK at 1 km distance from main campus building (Figure 1).

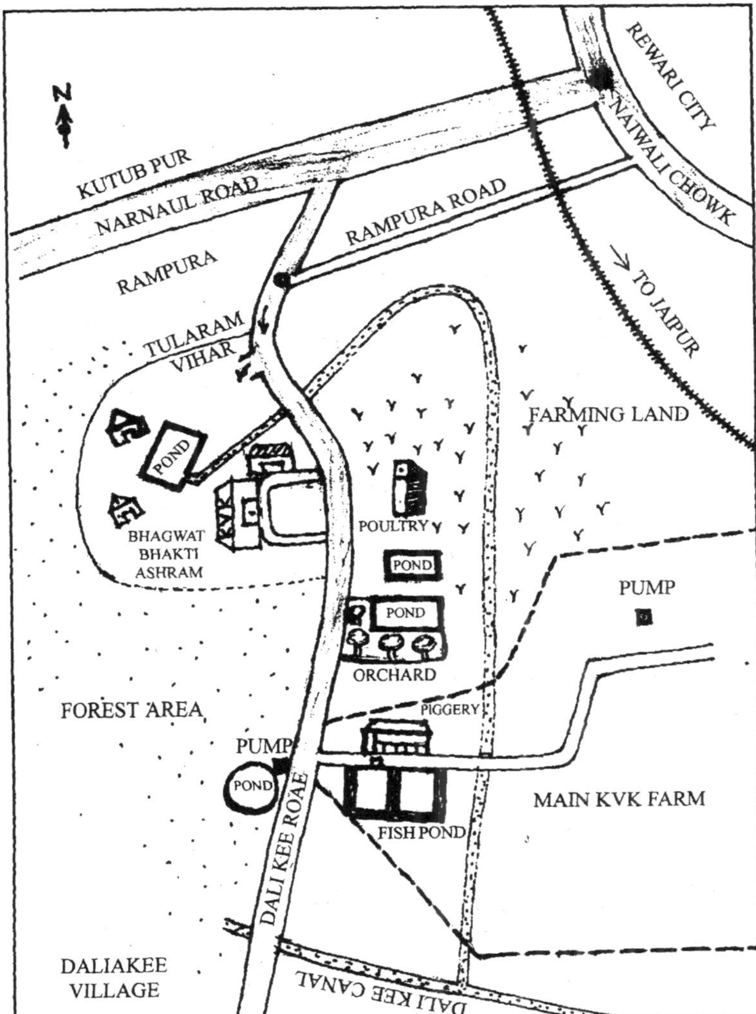

Figure 1: Sketch of Experimental Site

Experimental Unit

The experimental unit is comprising three components *viz.* fish pond, pig sties and azolla with integration of three commodities (fish, pig and azolla) for experiment and two commodities (fish, pig) as control (Figure 2 and Plate 1). The fish ponds are man made, rectangular in shape consisting the size dimension 40 × 30 × 1.5 m (area-0.12 ha) of each pond (Plate 4). Two earthen ponds of same size were selected for present study. These ponds are close to pig breeding unit. Ponds are five years old with regular practice of fish culture. Supply of water has been arranged through canal supply (Daliakee canal) and tube well of fresh water.

Plate 1: View of Experimental Unit

Piggery unit is located near on northern side of fish ponds, consisting asbestos roofed and cemented floor with 12 pig sties of different size (Plate 2). Fresh water of tube well was used for piggery unit. Pen no. 6 and no. 7 of the piggery unit were selected to keep the experimental pigs, each pen has 120 (10×12 sq feet) covered and 120sq. feet open area with proper arrangement of feeding and drinking trough. A channel of PVC pipe (diameter 4 inch and length 12 m) was fitted from open drainage of pen no. 6 to control pond (P1). An another channel of PVC pipe (length 6 m) was connected from pen no. 7 to azolla production-cum-stabilization pond and outlet (2.5 m) of this stabilization pond was connected to experimental fish pond (P2). A complete cemented rectangular pond (4×6 m^2) with depth of 60 cm was constructed for azolla production-

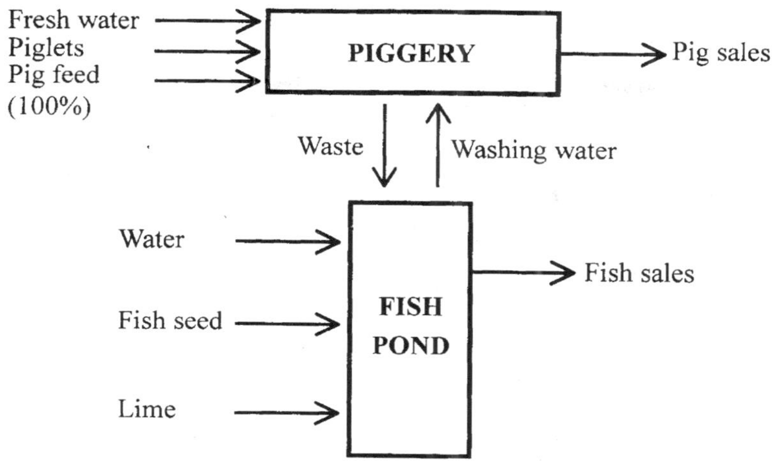

Pig-fish farming (control) unit

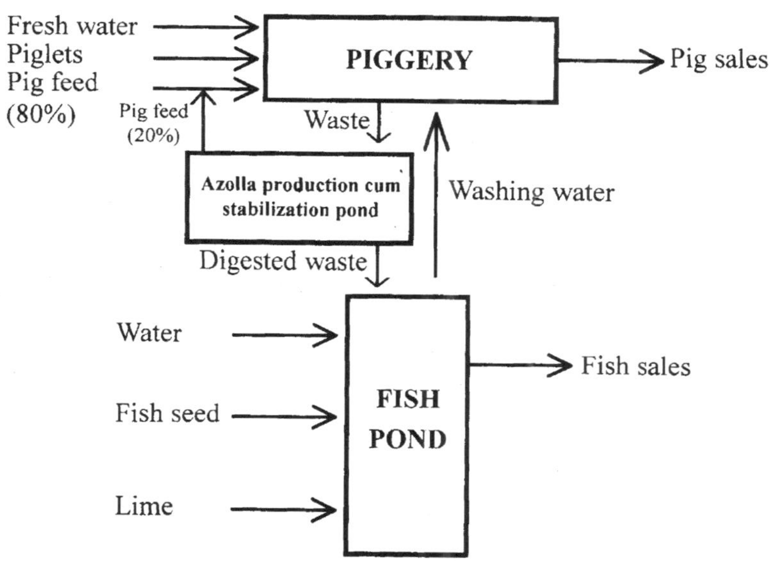

Azolla-pig-fish farming (experimental) unit

Figure 2: Design of Integrated Fish Farming System

Plate 2: View of Piggery Unit

cum-stabilization purpose, on the available space of dyke between piggery and fish ponds (Plate 3).

A built-in-shutter was provided in the drainage pipe of piggery unit to azolla production cum stabilization pond for regulating the flow of wastes.

Waste of the pig sty, pen no. 6 was directly drained into the pond (P1) daily, while waste of the pen no. 7 was drained to stabilization pond and it was drained weekly through L shaped outlet into the pond (P2) after harvesting azolla biomass. Excess water of the stabilization pond was also drained in intermediate days to maintain the water level (45 cm) in the digestion pond. After each drainage, stabilization pond was treated with lime. An another earthen azolla production pond was also constructed in same pattern to meet the excess requirement of azolla.

Large white Yorkshire breed was selected for experiment purpose (Plate 6). Weaned piglets of 2½ months age were carried out from breeding unit of KVK, Rewari. Six piglets were stocked in each pen during the month of October 2002. Rearing practice given by

Plate 3: *Azolla* Production-cum-stabilization Pond

Plate 4: Fish Production Pond

Sharma *et al.*, (1985) and Anon (1985, 1993) was followed. Piglets of male sex were castrated before stocking. Piglets group of pen no. 6 was fed with pigmash and other group of piglets of pen no. 7 was fed with azolla supplemented diet.

Composition of Pig Mash

Rice bran: 30 kg

Rice polish: 15 kg

Wheat bran: 27 kg

Maize broken: 10 kg

Mustered oil cake: 12 kg

Fish meal: 4 kg

Mineral mixture: 1 kg

Common salt: 1 kg

100 kg pig mash was prepared in each lot. Two times feeding (7 am and 5 pm) every day with different feeding rate according to the age have been done for each batch of piglets. Cleaning and washing of floor was done in morning at 6 am.

Feeding Rate (per day)

Age (Months)	*Quantity (kg)*
2½–4	1.0
4–6	1.5
6–8½	2.0

Azolla supplemented diet was prepared with 80 per cent pigmash and 20 per cent sun dried azolla. Harvested azolla biomass from digestion pond was used for this purpose.

First lot of the piglets was disposed and sold after 6months (April, 2003). The initial and final weight of the each animal was noted. A fresh lot of 5 piglets was stocked in each pen during the last week of April 2003. Initial weight of the all piglets were noted. Due to the high organic load, in ponds, the nos. of animals was reduced to 5 in second lot. The final weight of the animals of second lot were noted at the time of harvest (Oct. 2003).

Azolla production cum digestion pond was used for dual purpose *i.e.* production of azolla in diluted piggery effluent and digestion of piggery effluent using azolla. *Azolla pinnata* was selected for this purpose and carried out from IARI-New Delhi. Method of culture of this unit was followed as described by Goldman 1979; Fallow field, 1988 and Rao *et al.*, 1993. Planting material of azolla was inoculated @ 1 kg/m^2 and harvested weekly (Plate 7). Same quantity of the fresh innoculum was stocked for next week. Pond was managed with pig effluent manuring with out applying phosphatic fertilizer from outside. Carbofuron @ 1-2 ppm and bavistin 20 ppm were used for pests and fungal attack. Diluted extract of neem leaves was also sprayed for this purpose. Observations were taken for doubling time.

Harvested azolla was kept for sun drying and used for feed preparation of experimental (pen-7) animals. A separate azolla unit was prepared for excess azolla biomass production to fulfill the requirement. To study the suitable breed of azolla in existing climatic conditions, a trial was designed for different 4 varieties *viz. A. pinnata, A. rubra, A. microphylla, A. filiculoides.* Innoculum of all varieties was carried out from National Centre for conservation and utilization of blue green algae IARI–New Delhi. Experiments were conducted in 4 earthern race ways (1 × 4 × 0.6 m).

Composite fish farming of Indian major carps (*Catla catla, Cirrhinus mrigala* and *Labeo rohita*) and common carp (*Cyprinus carpio* var. *communis*) was adopted in both ponds (Plate 5). The package of practice described by Hichling (1962), Anon (1993) and Sharma *et al.* (1985) was followed. The eradication of unwanted fish was done by complete draining of the pond during the month of September. Ponds were filled with fresh canal water. An iron screen with net cloth was fitted in the canal near the inlet of the pond to prevent entry of wild fish in the ponds. Liming was done with slaked lime @ 200 kg/ha (24 kg in each pond). Ponds were stocked with advance sized fingerlings of selected species after 20 days of piglets stocking (1 Nov. 2002). Stocking was done @ 8000 fingerlings per hectare followed the stocking ratio of 4 species (4:3:2:1) composition given by Sharma *et al.*, (1985). In total stocking of 960 nos. of fish, catla 384, rohu 288, mrigal 192 and common carp 96 were stocked. Species wise initial average size was 74.5 g catla, 70.1 g rohu, 41.3 g mrigal and 137.2 g common carp.

Plate 5: Stocked Fish Species

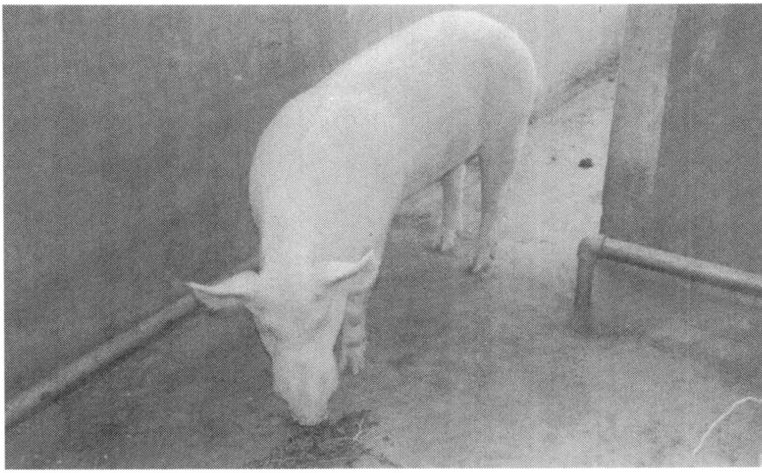

Plate 6: Adult Pig (Large White Yorkshire)

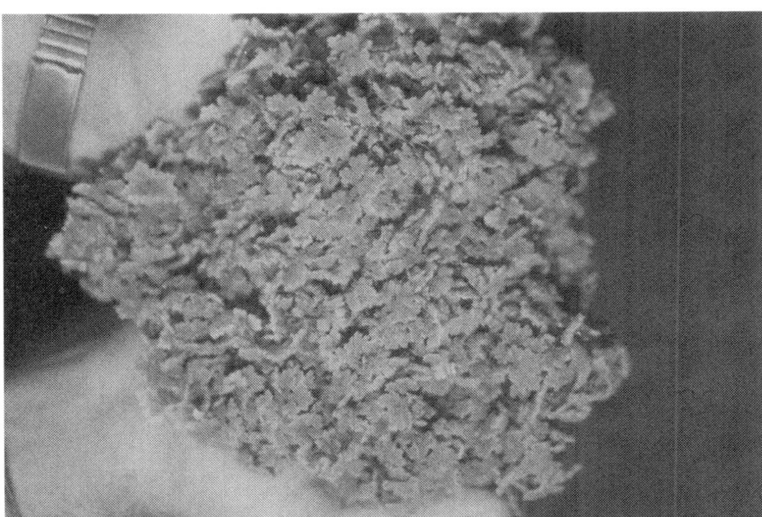

Plate 7: Fresh Innoculum of *Azolla* (*Azolla pinnata*)

Fingerlings were acclimatized for 4 days in nursery pond. Second dose of the lime @200 kg/ha was applied during the month of April at the time of piglet stocking of second lot. Final harvesting was conducted during October 2003 and 50 individuals of each species were observed for length and weight. Total harvested fishes were counted and weighed to calculate the mortality rate and final yield. Hence, in control unit, fish production was carried out through natural fish food organisms after recycling of fresh pig dung and in experimental unit, fish production was obtained through natural fish food organisms after recycling of digested pig dung of azolla fed pigs. Both units were designed and managed as integrated units without supplementary feeding with recycling of organic wastes to enhance the natural productivity. The difference in natural productivity and fish production between pig-fish and azolla-pig-fish culture unit was evaluated.

Chapter 4

Materials and Methods

The method adopted in investigating various biological aspect are based on standard evolved in different research laboratories both in India and abroad. Whenever necessary, the methodology was modified to suit the local conditions.

Meteorological Parameters

The meteorological data *i.e.* maximum and minimum air temperature and rainfall have been collected from the laboratory of KVK Rewari, which have all the facilities to record this meteorological observation on daily basis.

Chemical Properties of Pond Soil

The soil samples were collected from the pond bed with the help of a hand dredge. Samples were taken randomly from 5-6 different places. All the samples were air dried, ground to fine powder with a roller, mixed and sieved through a 2.0 mm sieve and kept in sealed plastic bags for further analysis.

Soil pH

To determine the hydrogen-ion-concentration, a mixture of soil and distilled water was prepared in the ratio of 1:2 and its pH was determined with pH meter (Black, 1965).

Table 1: Sampling and Analysis Schedule

Sl.No.	*Parameters*	*Intervals*
1.	**Soil**: pH, organic carbon, nitrogen, phosphorus.	Monthly
2.	**Water**: Water temperature, pH, dissolved oxygen, free carbon, alkalinity, salinity, TDS, ammonia, nitrite, nitrate, phosphate, and BOD_5.	Monthly
3.	**Phytoplankton**: Qualitative and quantitative estimation.	Monthly
4.	**Primary production**: Gross primary productivity. Net primary productivity, community resperation.	Monthly
5.	**Zooplankton**: Qualitative and quantitative estimation	Monthly
6.	**Benthos**: Qualitative and quantitative estimation Nitrogen in plankton/benthos	Monthly At end of experiment
7.	**Fish**: Length weight measurement, biochemical analysis–crude protein, crude fat, moisture and ash content.	Monthly At the end of experiment
8.	**Pig**: Estimation of growth in terms of live weight, feed analysis, FCR.	Monthly At the end of experiment
9.	**Pig excreta**: Pig excreta production, nitrogen, phosphorus, potassium, moisture.	December and March
10.	**Azolla**: Net production (fresh biomass) crude protein, crude fat, crude fibre, moisture	Weakly At the end of experiment

Organic Carbon

Organic carbon was determined by the method of Jackson (1958). 10 ml $K_2Cr_2O_7$ solution was thoroughly mixed with 1.0 g soil, kept in a conical flask. Subsequently, 20 ml of conc. H_2SO_4 was added to the mixture, and allowed to stand for 30 minutes. The volume was made up to 200 ml with distilled water, this was followed by the addition of 10 ml 85 per cent phosphoric acid and 1.0 ml of diphenylamine indicator. This mixture was titrated with 0.4 N Ferrous Ammonium Sulphate solution. Similar procedure was followed with a blank preparation as well. The percentage of organic carbon and organic matter, were calculated using the following equations:

$$\text{Per cent organic carbon (per cent C)} = \frac{3.951}{\cdot}\left(1\frac{T}{g\,S}\right)$$

Per cent organic matter = per cent C × 1.724

where,

g: Weight of sample in g

T: ml ferrous solution used with sample titration

S: ml ferrous solution used with blank titration

Nitrogen

30 ml of conc. H_2SO with1.0 g salicylic acid was added in Kjeldahl flask having 10 g soil. 5 g soidium thiosulphate was added and kept for half an hour. After adding 1.0g powdered copper sulphate, the mixture was digested on a gentle flame till the appearance of a clear blue colour. After cooling the solution was transfer to ammonia distillation flask, treated with 12 N NaOH using phenolphathalein as indicator. The distilation was done and the ammonia was collected in a 25 ml 0.1 N H_2SO_4 solution in a conical flask with few drops of methyl red indicator. About 150 ml of the distillate was collected and titrated with excess of 0.1 N H_2SO_4 with 0.1 N NaOH till the solution turns colourless.

Calculation

Total Nitrogen (%) = NaOH required × 0.0014 × 10

Phosphorus

The phosphorus content of soil samples was determined by the method followed by Olsen *et al.* (1954). Briefly, 20 ml of bicarbonate extract was added to 1.0 g of air dried soil sample kept in a flask, and was subjected to process of mechanical shaking for 30 minutes. The mixture was, subsequently filtered with whatmen no. 1 paper and 0.5 ml of molybdate reagent was added into the filtrate. The reaction mixture was then diluted with 20.0 ml of distilled water, which was followed by addition of few ml of stannous chloride solution. The optical density of this sample was determined at 660 nm in spectrophotometer and compared with a blank preparation.

$$\text{Per cent phosphorous} = \frac{\text{mg p/l in soil solution}}{50}$$

$$\text{Phosphorus (mg/100 g soil)} = \frac{\text{mg p/l in soil solution}}{50} \times 1000$$

Physico-Chemical Parameters of Pond Water

The water samples were collected in sampling bottles from each pond between 10-11 am for the estimation of pH, dissolved oxygen, free carbon dioxide, alkalinity, ammonia, nitrite, nitrate, phosphate, salinity, TDS and BOD_5. Dissolved oxygen was measured at pond itself while rest of the parameters were analysed in the laboratory. Water temperature was also measured in the ponds.

Water Temperature

Water temperature was recorded with the help of ordinary thermometer having range of 0-50°C, with mark upto 0.1°C.

Hydrogen-ion Activity (pH)

The pH of the water samples were measured electrometrically with a pH meter.

Dissolved Oxygen

The dissolved oxygen content of pond water was determined following Winkler titrimetric method (APHA, 1985).

Water samples were collected carefully in 100 ml glass stoppered sampling bottles within the water body to exclude air bubbles. 1.0 ml each of manganous sulphate and alkaline iodide reagents were added by means of 1 ml pipette, that was immersed to the bottom of the bottle and the pipette was slowly drawn out after addition. The stopper was replaced and the bottle was inverted 3-4 times for thorough mixing. The resultant flocculent was dissolved by adding 1 ml concentrated H_2SO_4, mixed and dissolved by gentle inversion. 50 ml of this solution was then transferred to an Erlenmeyer flask and 0.025 N $Na_2S_2O_3$ solution was added drop by drop till the colour turned to pale straw. 1 ml starch solution was added and titration continued to reach the end point, *i.e.* disappearance of the blue colour.

Calculation

Dissolved oxygen (mg/l) = ml of 0.025 N $Na_2S_2O_3$ used × 4

Free Carbon Dioxide

The free carbon dioxide was estimated by standard titrimetric method using phenolphthalein as an indicator (APHA, 1985).

100 ml of the sample was taken in a Nessler's tube and 5-10 drops of phenolphthalein indicator was added. The appearance of red colour indicated absence of free CO_2 in the sample. Sample remaining colourless indicated presence of free CO_2. 50 ml of this sample was titrated rapidly into the cylinder with standard alkali solution (N/44 NaOH) until a definite pink colour persisted.

Calculation

Free CO_2 was calculated by using the following expression:

$$CO_2\,(mg/l) = \frac{A \times N \times 44{,}000}{\text{ml of sample}}$$

where,

A: ml titration for sample

N: Normality of NaOH

Total Alkalinity

Alkalinity was assessed titrimetrically using phenolphthalein and methyl orange as indicators (APHA, 1985).

Phenolphthalein Alkalinity

5-6 drops of phenolphthalein indicator solution was added into 50 ml sample. If the sample remained colourless, phenolphthalein alkalinity was absent. The appearance of pink colour indicated the presence of phenolphathalein alkalinity. The sample was titrated with 0.02 N H_2SO_4 solution to a colourless end point.

Calculation

Phenolphthalein alkalinity was calculated by using the following expression.

$$\text{Phenolphthalein alkalinity } (CaCO_3 \text{ mg/l}) = \frac{A \times N \times 50{,}000}{\text{ml sample}}$$

where,

A: ml 0.02 N H_2SO_4 used for titration

N: Normality of acid (H_2SO_4)

Methyl Orange Alkalinity

5-6 drops of methyl orange indicator solution were added to 50 ml sample. The sample was titrated with standard 0.02 N sulphuric acid till yellow to faint orange colour developed.

Calculation

$$\text{Methyl orange alkalinity as mg/l } CaCO_3 = \frac{A \times N \times 50{,}000}{\text{ml of sample}}$$

where,

B: ml titrant used for sample to develop pink colour

N: Normality of acid (H_2SO_4)

Total alkalinity (mg/l): phenolphthalein alkalinity + methyl orange alkalinity

Salinity

The salinity of water was calculated from the EC values of water by following the relationship:

Salinity (ppm) = Electric conductivity (mm hos/cm) × 0.64

EC was determined with the help of an EC bridge, known as conducticity meter. Standard Potassium Chloride (KCl) was used to determined cell constant of the instrument. The conductance of water sample was measured with this instrument.

Calculation

EC (mm hos/cm) = conductivity × cell constant

Total Dissolved Solids

Total dissolved solids was estimated by standard method (APHA, 1985) 125 ml. Of water sample is filtered through a glass fibre filter placed in a funnel. The 100 ml of the filtrate is transferred to a pre weighed dish. The content are evaporated as for total solids.

Calculation

$$\text{TDS (mg/l)} = \frac{\text{R–D} \times 1000}{\text{Sample vol. (ml.)}}$$

where,

R: Weight (g) of the evaporating dish and residue

D: Weight (g) of the evaporating dish.

Ammonia-Nitrogen

Ammonical nitrogen was quantified by the direct nesselerization method (APHA, 1985). Accordingly, 50.0 ml of sample was nesselerized with 2.0 ml of nesselers reagent and the optical density of the sample was read with spectronic 20 at 410 nm.

Nitrite-Nitrogen

50 ml water sample was taken, 1.0 ml of EDTA followed by 1.0 ml of sulfanilic acid was added, mixed well, 1.0 ml 1-naphthyl amine hydrochloride and 1.0 ml sodium acetate was added. Blank was prepared in the same way by substituting the distilled water for the sample. Colour was measured after 10 minutes at 520 nm in 1.0 cm cell path. Calibration curve was prepared using suitable quantity of standard NO_2. Nitrated concentration in sample was calculated from the calibration curve.

Nitrate-Nitrogen

To determine the nitrate nitrogen phenol disulphonic acid method (APHA, 1985) was followed. 50.0 ml of water sample was evaporated in a silica crucible and the residue was dissolved in 1.0 ml of phenol disulphonic acid. Subsequently, the contents were diluted to 10 ml with ammonia free distilled water. Thereafter, the acidity of the reaction mixture was neutralized with 6.08 ml of strong ammonia solution and its optical density was determined with a spectronic 20 spectrophotometer at 410 nm. The concentration of

No_3-N was calculated from the calibration curve prepared from the known working solution of the nitrate in the range of 0.0 to 0.7 mg/l.

Phosphate-Phosphorus

For the determination of phosphorus content of pond water, 2.0 ml of ammonium molybdate reagent was added to 100 ml sample of water. Subsequently, 0.5 ml freshly prepared stannous chloride solution was also added to the sample whose optical density was measured using a spectronic 20 at 690 nm against a distilled water blank (APHA, 1985). Phosphorus concentration of the sample was calculated from the calibration curve made in the range of 0.0 to 0.05 mg/l.

Biochemical Oxygen Demand

The biochemical oxygen demand (BOD) is a measure of the amount of oxygen required by micro-organism to decompose the organic matter. Complete decomposition require a long time and hence a 5 day incubation period has been suggested for most practical purpose (APHA, 1971). The BOD value thus obtained was usually termed as BOD_5.

One liter of water sample was taken in a container and skaked thoroughly to increase its DO level. Two BOD bottles were filled with this sample water. Opaque bottle was incubated at 20°C. DO of the other bottle was determined following the method of DO estimation. DO of the incubated bottle was determined after the 5 days.

$$BOD_5\ (ppm) = DO\ (ppm) - DI\ (ppm)$$

where,

DO: Dissolved oxygen level in non incubated bottle.

DI: Dissolved oxygen level in incubated bottle.

Biological Estimations

Plankton Community Analysis

For qualitative analysis, the plankton samples were collected all along the pond filtering pond water through silk cloth plankton net of 25 cm diameter. Samples were preserved in 5 per cent formalin and were examined for the live specimens of plankton (phytoplankton and zooplankton) under a microscope. The identification was done with the help from Ward and Whipple (1959), Phillpose (1967), Pennak (1978), Tonapi (1980).

For the quantitative estimation of phytoplankton, 50 litres of water of each pond was filtered separately with the help of plankton net. The plankton samples thus collected were further centrifuged to concentrate the plankton population before counting. Enumeration of the phytoplankton was carried out using a bright line improved Haemocytometer. Four slides were prepared and counted for each sample and the results were calculated in terms of cells l^{-1} as follows (Stephens and Gillespie, 1976).

$$\text{No. of cells } l^{-1} = \frac{\text{No. of cells}}{0.9\ \text{mm}^3} \times \frac{1000\ \text{mm}^3}{1\ \text{cm}^3} \times \frac{1000\ \text{cm}^3}{1\ \text{litre}}$$

The final values were obtained by dividing the above values by the concentration factor of the initial samples. All the species were counted as a single unit whether colonial (*e.g. Microcystis*); single–celled (*e.g. Chlamydomonas*).

For the quantitative study of zooplankton, 50 litre sample of water was similarly filtered with the net and the numerical enumeration of zooplankton was carried out in a "sedgewick-rafter cell" adopting the procedure outlined by Welch (1948). A minimum of three sub samples were counted to determine statistical validity. The following equation was used to estimate the density of zooplankton.

$$n = \frac{(a.1000)\,c}{1}$$

where,

n: Number of plankters per litre

a: Average number of plankton in one small counting chamber of counting cell

c: Volume of concentrate in ml

1: Volume of water filtered in litres.

Total volume of plankton was taken to note the reading of graduated test tube of plankton net.

Primary Production of Phytoplankton

The primary production of phytoplankton was measured by "light and dark bottle method" (Gaarder and Gran, 1927). One pair of light and dark bottles containing water samples were suspended *in situ* with the help of floats for a period of 6 hrs. DO of the samples was measured before and after each experiment using Winkler's method. The gross oxygen production (GOP) were determined from the changes in the oxygen concentration in light and dark bottles. The productivity was calculated in terms of carbon by multiplying GOP with a factor 0.375 (Westlake, 1963; Sreenivasan, 1964). The following equations were used for calculating gross production.

$$\text{Gross production (g C m}^{-3}\text{ hr}^{-1}) = \frac{\text{Lo - Do} \times 0.375 \times 1000}{\text{T} \times \text{P.Q.}}$$

$$\text{Net production (g C m}^{-3}\text{ hr}^{-1}) = \frac{\text{Lo - Io} \times 0.375 \times 1000}{\text{T} \times \text{P.Q.}}$$

$$\text{Community respiration (g C m}^{-3}\text{ hr}^{-1}) = \frac{\text{Io - Do} \times 0.375 \times 1000}{\text{T} \times \text{P.Q.}}$$

where,

Lo: Dissolved oxygen in light bottle after exposure

Do: Dissolved oxygen in dark bottle after exposure

Io: Initial oxygen concentration

PQ: is the photosynthetic quotient which is assumed at 1.2 (Fogg, 1963)

T: Time (hrs.)

Benthic Macroinvertebrate Community

Samples for macrobenthic faunal analysis were collected from different places of pond bottom with the help of a hand dredge (25

cm^2) and kept in polythene bags for further investigations in laboratory. These samples were first sieved alongwith water through brass strainer (0.5 mm × 2.0 mm pores). The benthic invertebrates were collected from the sieve with a forcep and kept in a wide mouth bottle containing 5 per cent formalin (70 per cent ethyl alcohol was used for organisms having calcareous shells). The organisms were identified according to the procedure described by Pennak (1978) and Tonapi (1980). The animals were counted groupwise and their density was calculated as follows:

$$\text{Density (no. m}^{-2}\text{)} = \frac{N}{a} \times 10^4$$

where,

N: Average number of organisms per sample.

a: Area of the sampler (cm^{-2})

Wet weight was taken on digital weighing machine. Dry weight was recorded after drying at 80°C for 8 hrs in oven.

Nitrogen Analysis in Plankton, Benthos

100 mg sample of each of the above biotic communities was taken for analysis. The estimations were done following the methods used to analyse nitrogen in soil.

Fish Growth

The growth rate of fish was measured in terms of gain in length and weight. Sampling was done at monthly intervals by netting out 10 fishes of each species from each pond and length and weight of each species measured separately. Observations for disease or infection were also made for each fish.

After netting the fishes were kept in a conditioning hapa. Weight and length were measured with the help of single pan balance and measuring board, respectively. Fishes were screened through hand lens for occurrence of any infection. Fishes were treated with the solution of $KMnO_4$ (5 mg/lit for 10 min.) before releasing in the ponds. Total fishes were harvested at the end of experiment. Each harvested fish was measured and weigh to quantify the production. Specific growth rate is estimated by formula given below:

$$SGR = \frac{\text{Loge final weight–loge initial weight}}{\text{Time}} \times 100$$

The length weight relationship of the experimental fish is worked out as per cube law given by Le Gren (1951) as follows:

$$W = CL^3$$

where,

W: Weight of fish

C: Constant factor

L: Length of fish.

The logarithmic transformation of the above formula is

$$\text{Low } W = \log a + n \log L$$

The condition factor (K) of the fishes was calculated as per formula given below:

$$K = \frac{W \times 100}{L^3}$$

where,

W: Weight of the fish (g)

L: Length of the fish (cm)

$$\text{Survival rate} = \frac{\text{Number of fishes recovered}}{\text{Number of fishes stocked}} \times 100$$

Bio-Chemical Analysis of Fish

Fish was analysed for crude protein, crude fat, ash content and moisture content.

Crude Protein

Micro-Kjeldahl method was followed for the estimation of total nitrogen. 2 gm dried sample was digested with 20-25 ml of concentrated H_2SO_4 in a Kjeldahl flask. A spoon of catalyst (mixture of copper sulphate and sodium sulphate in 5 : 95 ratio) was added to hasten the process of digestion. The flask was then kept on digestion bench and heated until the sample turned green and it was made sure that no trace of black particle remained in the sample. After cooling, the digested sample was transferred into 250 ml volumetric flask and volume was made up to the mark. Aliquot sample of 10 ml was taken into micro Kjeldahl distillation set for distillation. The liberated NH_3, upon addition of 40 per cent NaOH was absorbed by boric acid solution containing Toshiro's indictor (8 mg methyl red and 20 mg bromocresol green dissolved in 100 ml methyl alcohol, 10 ml of this indicator added in 1 liter of 2 per cent boric acid solution) and titrated against N/70 H_2SO_4. Total protein was calculated as follows:

Total protein (g protein/100 g sample) = A × 0.0002 × 25 × 50 × 6.25 where, 'A' represents biurette reading of N/70 H_2SO_4.

Ether Extract Estimation (Crude Fat)

About 3 g sample was taken into a thimble. Thimble was kept in a beaker and dried in oven for about 6 hours at 100°C. Then the thimble was transferred to the soxhlet extraction apparatus. Petroleum ether (60-80°C) was taken in a dried, weighed extraction flask of the extraction apparatus and was fitted into the stem of the extraction apparatus. The apparatus was allowed to run for about 4-6 hours. The flask was removed after recovering the ether of the flask and the last traces of solvent evaporated on a water bath. After drying, the extraction flask was kept in drying air oven at 100°C for 30 minutes. Then it was transferred to a desicator and weighed after the flask had cooled to room temperature.

Fat percentage was calculated as follows:

$$\text{Crude fat (per cent)} = \frac{A_2 - A_1}{A} \times 100$$

where,

A_1: Weight of empty extraction flask

A_2: Weight of extraction flask with material

A: Weight of sample

Moisture Content

Ten grams of feed/fish sample was accurately weighed in a chemical balance in a dry clean moisture cup and kept in hot air oven at 100°C for 16-18 hours. The moisture cup was removed from oven, cooled in a desicator and weighted. Weight loss was taken as moisture content of the sample and per cent moisture was calculated as follows:

$$\text{Moisture (per cent)} = \frac{\text{B - C}}{\text{B - A}} \times 100$$

where,

A: Weight of moisture cup

B: Weight of moisture cup + wet sample

C: Weight of moisture cup + dried sample

Dry matter was calculated by using following expression:

$$\text{Dry matter (per cent)} = \frac{\text{C - A}}{\text{B - A}} \times 100$$

Ash Content

Dry samples obtained after moisture estimation were taken in a dry, clean silica basin and heated until samples were well carbonized and silica basin was transferred to a muffle furnace at a temperature of 550°C. Silica basin was removed along with white ash and after cooling in a desicator final weight taken. Per cent ash was calculated by using following expression:

$$\text{Ash (per cent)} = \frac{\text{C - D}}{\text{E - D}} \times 100$$

where,

C: Weight of silica basin + white ash

D: Weight of silica basin

E: Weight of silica basin + dried material

Pig Growth and Yield

Initial live weight of all piglets (2½ months old) were taken and calculated average initial weight of piglet in each batch of each lot. Final live weight was taken at the time of sale. Net gain in weight was calculating as follows:

Net gain in wt. (kg) = Average final weight

$$\text{Feed conversion ratio (FCR)} = \frac{\text{Feed intake (kg)}}{\text{Net gain in weight (kg)}}$$

Feed samples of both groups of animals were collected and analysed for crude protein, crude fat and crude fibre (crude protein and crude fat was estimated following the methods for biochemical analysis of fish).

Crude Fibre

Crude fibre was estimated following the method A.O.A.C. (1980).

$$C\ (\%) = (\text{loss in wt. On ignition–loss in wt. of asbestos blank}) \times 100/\text{wt sample.}$$

Animals of three age group (3, 4, 8 months) were kept in a separate pen and samples of their waste (dung + urine) were collected for 24 hrs. calculation was done on average basis and computed waste/animal/day. Production of waste by individual animal in 6 months calculated as follow:

$$\text{Waste (in 6 months) kg} = \text{waste/ animal/ day} \times 6 \times 30$$

Production and Analysis of Pig Excreta

Animal wastes (dung + urine) were collected for 7 days and dried to expose bright sun light. Three sub samples of 100 g each have been analysed for proximate nutrients percentage. Samples of azolla treated peggery wastes were also analyzed. All samples were collected in plastic buckets and weighed accordingly to estimate moisture content. Sun dried samples were packed in polythen bags for laboratory work. Samples of the whole day wastes were collected three times during the rearing period, when animals have been attained the age of 3, 4 and 8 months. Average of the weight of all samples was calculated to estimate the average quantity of waste voided by single animal in a day. The total quantity of waste recycled in the pond was computed by average quantity, numbers of rearing days and numbers of stocking animals. On the basis of proximate nutrients composition, the quantity of each nutrients from recycled wastes was calculated.

Nitrogen, phosphorus (followed the method of soil analysis), potassium was estimated following standard method.

Potassium

Potassium was estimated by flame photometric procedure (Black, 1965). 5.0 g dried sample was extracted with 25 ml ammonium acetate solution for 5 minutes and filtered with whatman no. 42 paper. The mixture was then analysed with flame photometer previously adjusted to 0-100 scale using 0-100 mg ml^{-1} of standard respectively.

Azolla Production

During the month of February, data were recorded for estimating growth and yield of azolla. 10 kg of innoculum was stocked and harvested after 7 days. After draining of digested slurry and cleaning of pond, a fresh innoculum was stocked. The accurate weight of harvested azolla was noted upto 3 week and computed average production and doubling period of azolla.

Analysis of *Azolla*

Samples of oven dried azolla were analyzed for crude protein, crude fat, crude fibre and moisture content as per method adopted for biochemical analysis of fish and analysis of pig feed.

Economic Analysis

Comparative production economics of pig-fish unit (control) and azolla-pig-fish unit has been worked out for total production cost and total return. Net return with percent return to variable cast has been computed from those data to asses the economic feasibility of the experimental unit over control unit.

Statistical Analysis

Soil and water quality parameters, phytoplankton, zooplankton, macrobenthic invertebrate fauna and fish/pig/azolla growth obtained in different experimental ponds were statistically tested for significant variations using students 't' test. Correlation among the different factors affecting the nutrient exchange was calculated. Correlation matrix was also calculated.

Chapter 5

Observation

A comparative study on the integrated pig-fish culture and azolla-pig-fish culture with special reference to pond productivity and fish production was carried out from November 2002 to October 2003. The results of different parameters are briefly described as follows:

Meteorological Parameters

Experiment was conducted at Krishi Vigyan Kendra, Rampura, Rewari. This location is particularly situated in southern Haryana, affected by semi-arid climate of Rajasthan. The minimum air temperature (3.5°C) was recorded in January and maximum (48.5°C) was recorded in the month of June with wide fluctuation during the experiment. The total rainfall (88.2 mm) was recorded during the experimental period with maximum (56.5 mm) in September (Table 2).

Chemical Characteristics of Pond Soil

Observation recorded on various chemical properties of soil are presented in Tables 3-6 and Figures 3-6.

pH remained slightly alkaline in both ponds through out the experimental period. The pH value of pond sediments varied between 7.4-7.6 and 7.3-7.6 in experimental and control pond, respectively. In both ponds, maximum and minimum values of pH were recorded in the beginning and end of experiment, respectively. Mean values in experimental (7.5±0.09) and control pond (7.4±0.12) are non-significantly different ($P<0.05$) (Table 3, 4 and Figure 3).

Table 2: Meteorological Data of Rewari During Experimental Period (Nov. 2002 to Oct. 2003).

Month	*Min. Temp. (°C)*	*Max. Temp. (°C)*	*Rainfall (mm)*
Nov	8.0	29.0	-
Dec	5.0	26.0	-
Jan	3.5	20.5	4.0
Feb	4.5	26.0	6.5
Mar	9.0	30.5	1.2
Apr	25.0	38.0	-
May	26.0	46.0	4.5
Jun	30.5	48.5	55
Jul	22.5	42.5	7.5
Aug	21.5	36.0	2.5
Sep	22.0	37.0	56.5
Oct	20.0	36.5	-

Source: KVK, Rampura, Rewari.

Organic carbon content ranged from 0.54-1.385 per cent in experimental pond and 0.55-1.438 per cent in control pond. The concentration of organic matter was maximum in the month of February in control pond, while it was maximum in the month of May (1.410 per cent) in experimental pond. In general, it was maximum towards the end of experiment in both ponds. Non significant difference was observed in average values of experiment 1.22±0.26 and control 1.21±0.25 unit (Tables 3, 4 and Figure 4).

Nitrogen concentration ranged from 30.6-66.0 mg/100g soil in experimental pond and 29.47-59.37 mg/100 soil in control pond. Maximum concentration was recorded during summer months. On an average, maximum concentration (66.0 mg/100g soil) was recorded in experimental pond. Mean values in experimental (49.14 mg/100g soil) and control (48.97 mg/100g soil) are non significantly different ($P<0.05$) (Tables 3,4 and Figure 5).

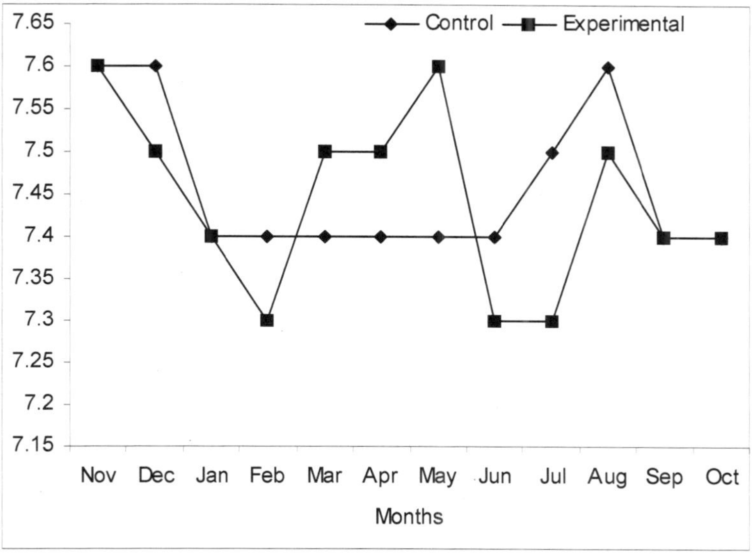

Figure 3: Seasonal Variation in Soil pH

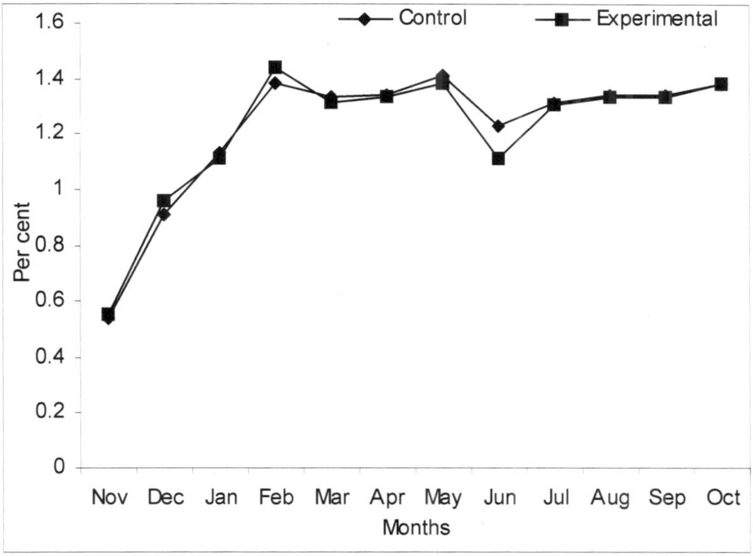

Figure 4: Seasonal Variation in Organic Carbon of Pond Soil

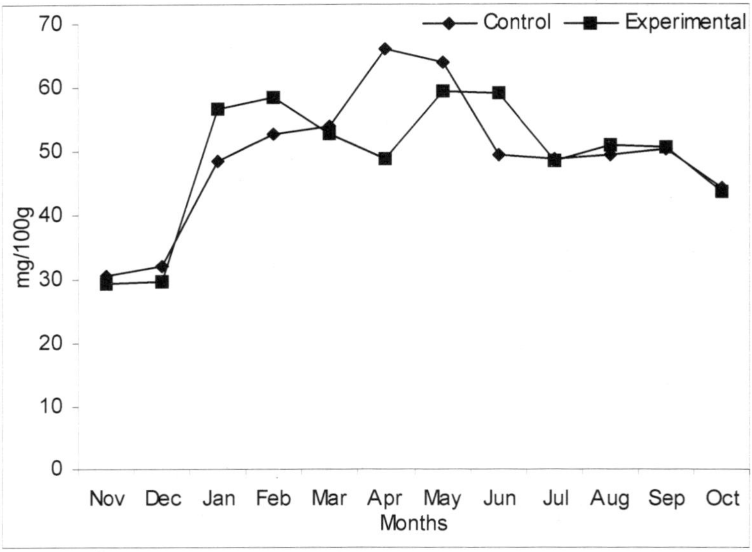

Figure 5: Seasonal Variation in Available Nitrogen Content of Pond Soil

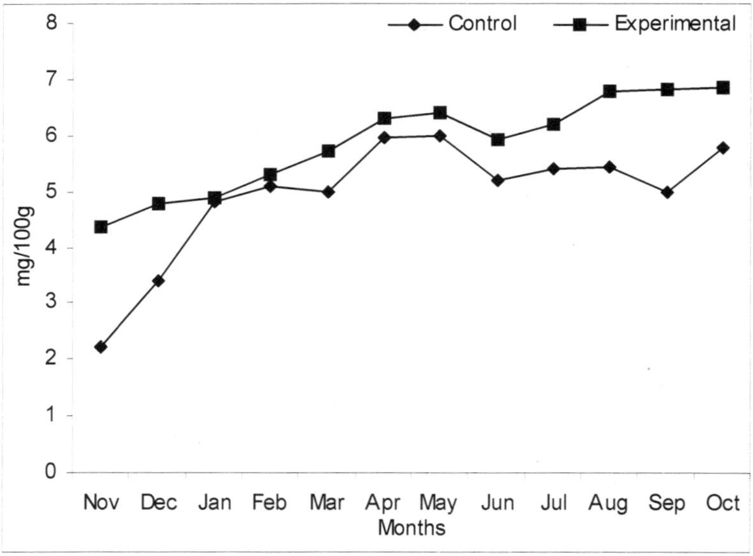

Figure 6: Seasonal Variation in Phosphorus Content of Pond Soil

Table 3: Temporal Variation in Chemical Characteristic of Experimental (E) and Control (C) Pond Soil

Month	*pH*		*Organic Carbon (%)*		*Available Nitrogen (mg/100g soil)*		*Available Phosphorus (mg/100g soil)*	
	E	*C*	*E*	*C*	*E*	*C*	*E*	*C*
Nov	7.6	7.6	0.540	0.550	30.60	29.47	2.21	4.38
Dec	7.6	7.5	0.910	0.954	32.00	29.60	3.38	4.79
Jan	7.4	7.4	1.132	1.112	48.46	56.56	4.80	4.90
Feb	7.4	7.3	1.385	1.438	52.85	58.38	5.10	5.31
Mar	7.4	7.5	1.332	1.311	54.00	52.58	4.98	5.72
Apr	7.4	7.5	1.338	1.337	66.00	48.88	5.96	6.31
May	7.4	7.6	1.410	1.380	64.00	59.37	5.99	6.40
Jun	7.4	7.3	1.230	1.112	49.25	59.20	5.20	5.92
Jul	7.5	7.3	1.314	1.310	48.64	48.60	5.40	6.20
Aug	7.6	7.5	1.340	1.332	49.44	51.00	5.44	6.80
Sep	7.4	7.4	1.338	1.334	50.30	50.46	4.99	6.82
Oct	7.4	7.4	1.385	1.380	44.12	43.50	5.80	6.84

Table 4: Mean Values (± SD) of Chemical Characteristics of Sediment in Experimental (E) and Control (C) Pond

Parameters	*Pond*	
	E	*C*
pH	$7.5^{a} \pm 0.09$	$7.4^{b} \pm 0.12$
Organic carbon (per cent)	$1.22^{a} \pm 0.26$	$1.21^{b} \pm 0.25$
Available nitrogen (mg/100g soil)	$49.14^{a} \pm 10.48$	$48.97^{b} \pm 10.29$
Available phosphorus (mg/100g soil)	$4.94^{a} \pm 1.10$	$5.87^{a} \pm 0.85$

Figures having same superscript in each row are significantly different ($P < 0.05$).

There is a gradual increase in phosphorus content from beginning to end of experiment. It was in the range of 2.21-5.99 mg/100g soil) in experimental and 4.38-6.84 mg/100g soil in control pond. The average (5.87 mg/100g soil) value of phosphorus in control pond soil was higher than in experimental pond (4.94 mg/100g soil), which was significantly different ($P<0.05$).

In experimental pond soil, pH showed significant negative correlation with organic carbon (–0.653), available nitrogen (–0.692), available phosphorus (–0.651) and GPP (–0.652). Organic carbon showed significant positive correlation with available nitrogen (0.809), available phosphorus (0.953) and GPP (0.601). Available nitrogen showed positive correlation with available phosphorus (0.857) and GPP (0.797). Available phosphorus showed positive correlation with GPP (0.709) at 1 per cent and 5 per cent level. Almost similar trend was observed in control pond (Tables 5, 6).

Table 5: Correlation Matrix of Different Chemical Characteristics of Sediment of Experimental Pond and GPP

	pH	*Org. Carbon*	*Average N_2*	*Average PO_4*	*GPP*
pH	1	–0.653*	–0.692*	–0.651*	-0.652*
Org. Carbon		1	0.809**	0.953**	0.601*
Average N_2			1	0.857**	0.797**
Average PO_4				1	0.709**

** Significant at 1 per cent level; * Significant at 5 per cent level.

Table 6: Correlation Matrix of Different Chemical Characteristics of Sediment of Control Pond and GPP

	pH	*Org. Carbon*	*Average N_2*	*Average PO_4*	*GPP*
pH	1	-0.376	-0.418	0.210	-0.156
Org. Carbon		1	0.696*	0.588*	0.757**
Average N_2			1	0.436	0.423
Average PO_4				1	0.639*

** Significant at 1 per cent level; * Significant at 5 per cent level.

Physico-Chemical Characteristics of Pond Water

Average value of various physico-chemical properties of water recorded in experimental and control pond are presented in Tables 7-11. Figures 7-18, depict the seasonal fluctuations of these properties.

Surface water temperature of experimental and control ponds fluctuated between 14.7-32.0°C. Minimum values were observed in

Table 7: Temporal Variation of Different Abiotic Parameters in Experimental Pond Water

Months	*Water Temp. (°C)*	*pH*	*DO (mg/l)*	*Co_2 (mg/l)*	*Total Alkalinity (mg/l)*	*Salinity (ppt)*	*TDS (mg/l)*	*NH_4-N (mg/l)*	*NO_2-N (mg/l)*	*NO_3-N (mg/l)*	*Phosphate (mg/l)*	*BOD_5 (mg/l)*
Nov	20.0	8.0	4.6	0.0	91	0.6	180	0.010	0.017	0.07	0.12	5.0
Dec	16.4	7.5	4.4	7.2	151	0.4	208	0.028	0.018	0.31	0.14	7.5
Jan	14.9	7.7	5.2	9.2	148	0.8	198	0.052	0.040	0.62	0.34	8.0
Feb	18.9	7.3	5.0	8.0	172	1.0	220	0.128	0.045	0.88	0.32	11.5
Mar	20.8	7.2	4.9	5.8	174	1.4	280	0.146	0.055	0.96	0.28	1.0
Apr	26.5	6.9	6.2	0.9	177	1.8	252	0.218	0.046	0.86	0.28	0.9
May	31.8	7.2	6	2.2	172	1.8	260	0.226	0.040	1.22	0.36	0.8
Jun	31.5	7.8	4.8	2.6	174	2.2	215	0.241	0.038	1.21	0.20	5.0
Jul	28.2	7.9	5.2	5.2	134	0.8	220	0.222	0.032	0.68	0.18	7.5
Aug	27.0	7.8	5.4	5.8	151	0.8	280	0.123	0.038	0.64	0.18	8.5
Sep	27.0	7.6	6.8	8.0	148	0.6	298	0.118	0.045	0.84	0.22	10.1
Oct	26.5	7.6	6.4	8.0	172	0.6	292	0.128	0.020	0.98	0.18	8.0

Table 8: Temporal Variation of Different Abiotic Parameters in Control Pond Water

Months	*Water Temp. (°C)*	*pH*	*DO (mg/l)*	*Co_2 (mg/l)*	*Total Alkalinity (mg/l)*	*Salinity (ppt)*	*TDS (mg/l)*	*NH_4-N (mg/l)*	*NO_2-N (mg/l)*	*NO_3-N (mg/l)*	*Phosphate (mg/l)*	*BOD_5 (mg/l)*
Nov	20.1	7.8	4.4	0.0	98	0.8	184	0.012	0.018	0.10	0.18	7.5
Dec	14.7	7.5	4.2	8.0	126	0.8	211	0.022	0.021	0.32	0.20	12.0
Jan	14.87	6.9	5.0	11.0	134	0.6	190	0.058	0.060	0.58	0.48	17.0
Feb	18.0	6.8	4.6	11.4	164	0.8	290	0.126	0.045	0.46	0.29	20.5
Mar	21.0	6.9	4.8	11.1	168	1.2	280	0.116	0.068	0.82	0.34	20.0
Apr	26.4	7.7	5.4	1.6	160	1.8	232	0.118	0.058	0.86	0.48	15.5
May	32.0	7.9	4.8	4.6	168	1.6	260	0.116	0.040	0.94	0.48	17.0
Jun	31.5	7.8	4.6	2.4	160	1.8	208	0.118	0.038	0.62	0.26	14.0
Jul	29.0	7.7	4.8	4.6	132	1.0	232	0.220	0.058	0.42	0.16	18.0
Aug	27.2	7.5	5.0	8.0	126	0.8	315	0.122	0.062	0.40	0.14	19.5
Sep	27.1	7.5	5.2	9.8	164	0.8	319	0.118	0.058	0.88	0.28	20.0
Oct	*26.2*	*7.3*	*5.2*	*10.0*	*160*	*0.4*	*266*	*0.118*	*0.060*	*0.38*	*0.32*	*18.0*

the month of December and January and maximum during May and June. The pattern of fluctuations was same in both ponds. Mean value in experimental (24.13±5.7) and control pond (24.01±6.07) are known significantly different (Tables 7, 9 and Figure 7).

pH of pond water varied from 6.9–8.0 (mean value 7.54±0.33) in experimental and 6.8–7.9 (mean value 7.44±0.39) in control pond. Maximum values were observed during November and May in experimental and control ponds, respectively. Minimum values were observed during the month of April and February in experimental and control ponds, respectively (Tables 7, 9 and Figure 8).

Dissolved oxygen concentration showed two peaks during the experimental duration. First in the month of April-May and second in the month of September-October. Highest value (6.8 mg/l.) was recorded in experimental pond, during the month of September, while it was minimum (4.4 mg/l) during the month of December. In control pond, maximum value (5.2 mg/l) was recorded during September and minimum (4.2 mg/l) during the month of December. Mean values in experimental (5.41±0.77mg/l) and control pond (4.83±0.35 mg/l) are significantly different (Tables 7, 9 and Figure 9).

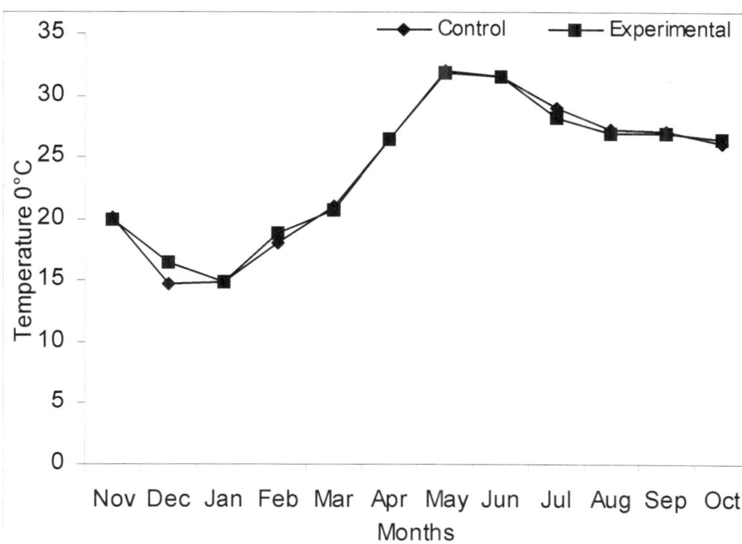

Figure 7: Seasonal Variation in Pond Water Temperature

Table 9: Mean Values (± SD) of Physico-chemical Characteristics of Water in Experimental (E) and Control (C) Pond

Parameters	Pond	
	E	*C*
Water temperature (°C)	24.13[a] ± 5.70	24.01[b] ± 6.07
pH	7.54[a] ± 0.33	7.44[b] ± 0.39
Dissolved oxygen (mg/l)	5.41[a] ± 0.77	4.83[a] ± 0.35
Free carbon dioxide (mg/l)	5.72[a] ± 2.75	7.5[b] ± 3.61
Total alkalinity (mg/l)	155.33[b] ± 24.75	146.67[a] ± 22.65
Salinity (ppt)	1.07[a] ± 0.59	1.03[b] ± 0.47
Total dissolved solids (mg/l)	241.92[a] ± 39.98	248.92[b] ± 46.43
Ammonia-nitrogen (mg/l)	0.14[a] ± 0.08	0.11[b] ± 0.06
Nitrite-nitrogen (mg/l)	0.04[a] ± 0.01	0.05[a] ± 0.02
Nitrate-nitrogen (mg/l)	0.78[a] ± 0.32	0.57[b] ± 0.26
Phosphate-phosphorus (mg/l)	0.23[a] ± 0.08	0.30[a] ± 0.13
BOD_5 (mg/l)	6.15[a] ± 3.64	16.58[a] ± 3.85

Figures having same superscript in each row are significantly different ($P < 0.05$)

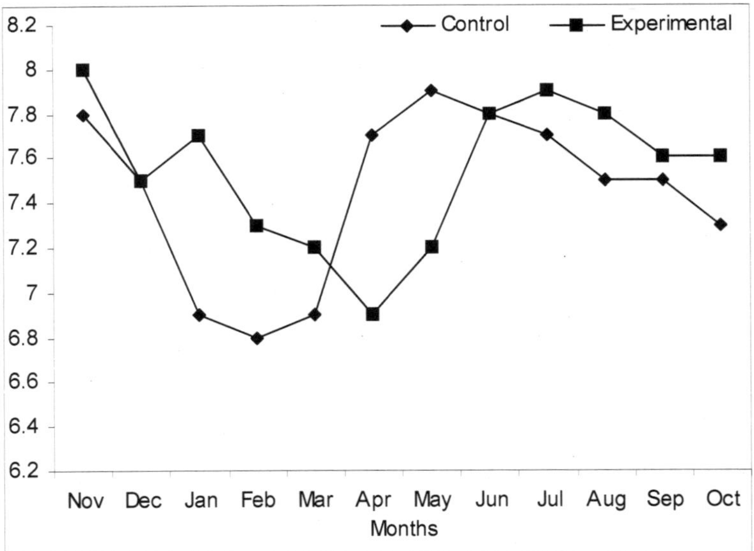

Figure 8: Seasonal Variation in Concentration of Pond Water pH

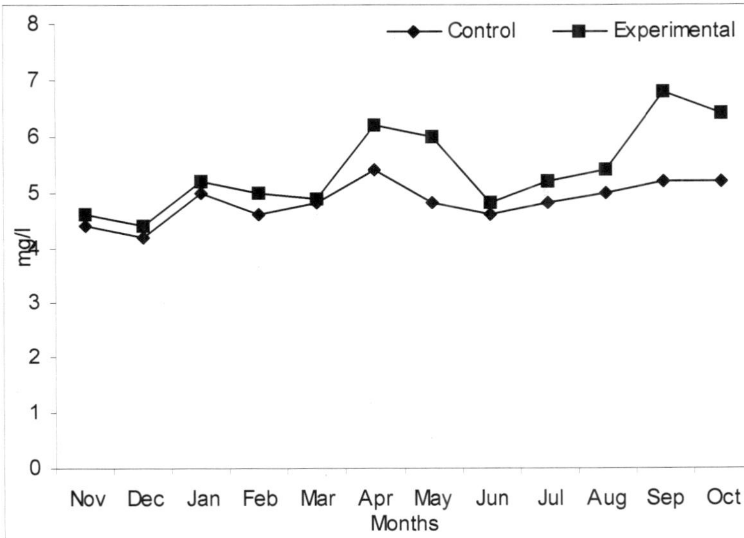

Figure 9: Seasonal Variation in Concentration of Dissolved Oxygen

Free carbon dioxide content in experimental pond was found in the range of 0.0-9.2 mg/l in experimental and 0.0-11.4 mg/l in control pond during the course of investigation. Higher values were recorded during the month of January and February in experimental and control pond, respectively. It was nil in the beginning and also found minimum during April-May. Fluctuation pattern was same in both ponds. Mean values in experimental (5.72±2.75 mg/l) and control pond (7.5±3.61 mg/l) are significantly different (Tables 7, 9 and Figure 10).

Gradual increasing trend was observed in the values of total alkalinity in both ponds, which was returned to down during July-August and increased again in the end of experiment. Average values were slightly higher in experimental pond. It was in the range of 98-168 mg/l and 91-177 mg/l. in control and experimental pond, respectively. Mean values in experimental (155.33±24.75 mg/l) and control pond (146.67±22.65 mg/l) are significantly different (Tables 7, 9 and Figure 11).

One peak in the values of salinity was observed during summer months. These values were in the range of 0.4-2.2 (1.07±0.59) ppt

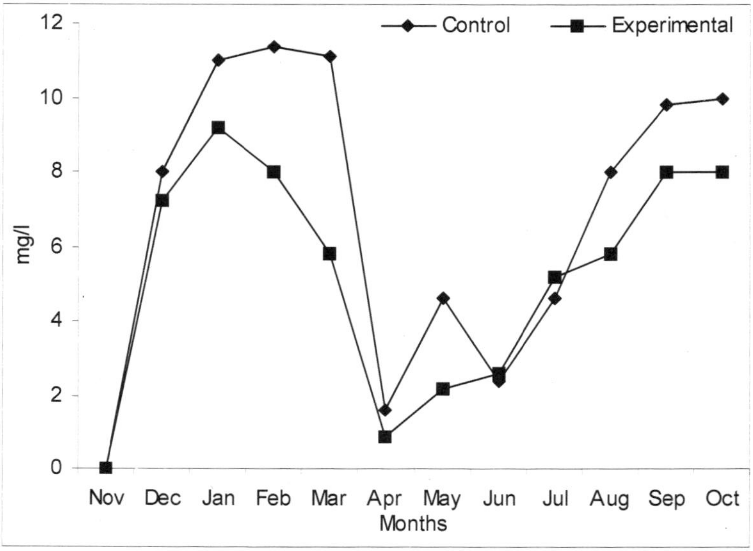

Figure 10: Seasonal Variation in Concentration of Free Carbon Dioxide

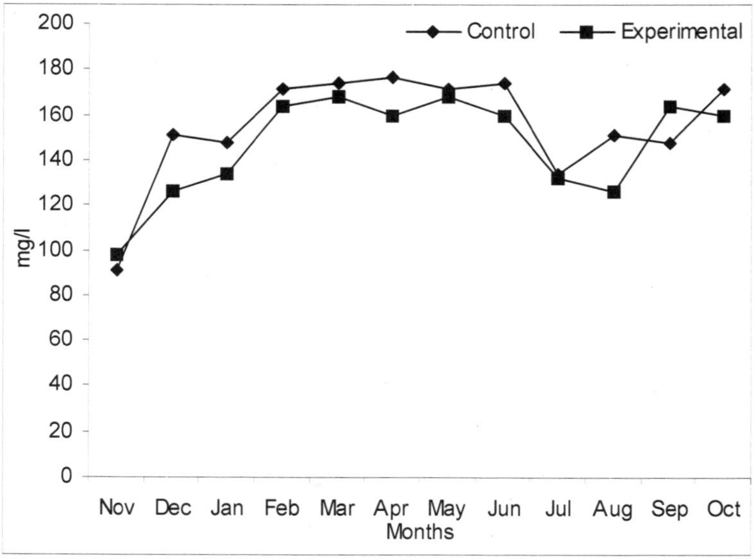

Figure 11: Seasonal Variation in Concentration of Total Alkalinity

and 0.4-1.8 (1.03±0.47) ppt in experimental and control pond, respectively. Highest value was observed during June and minimum in the month of December and January in both ponds. Mean values in experimental and control pond are non-significantly different (Tables 7, 9 and Figure 12).

Continuous fluctuation was observed in the values of total dissolved solids. It was observed in the range of 180-298 mg/l and 184-319 mg/l in experimental and control pond, respectively. Highest values were observed during September and minimum in November in both ponds. Mean values in experimental (241.02±39.98 mg/l) and control pond (248.92±46.43 mg/l) are non-significantly different (Tables 7, 9 and Figure 13).

The concentration of NH_3-N was higher in experimental pond with the range of 0.01-0.24 mg/l with highest value in summer months. While it was in the range of 0.02-0.220 in control pond. Ammonia concentration increased with the progress of the experiment and reaching peak values in the month of August. The pattern of fluctuation was almost similar in both the ponds. Mean

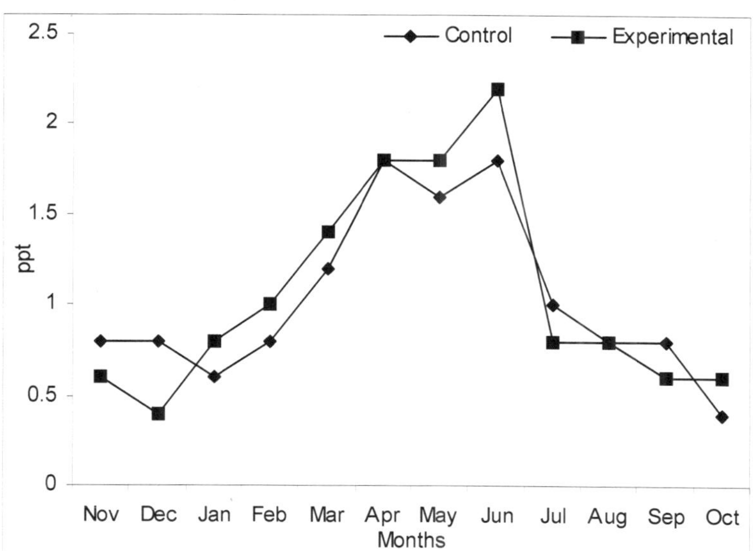

Figure 12: Seasonal Variation in Concentration of Salinity

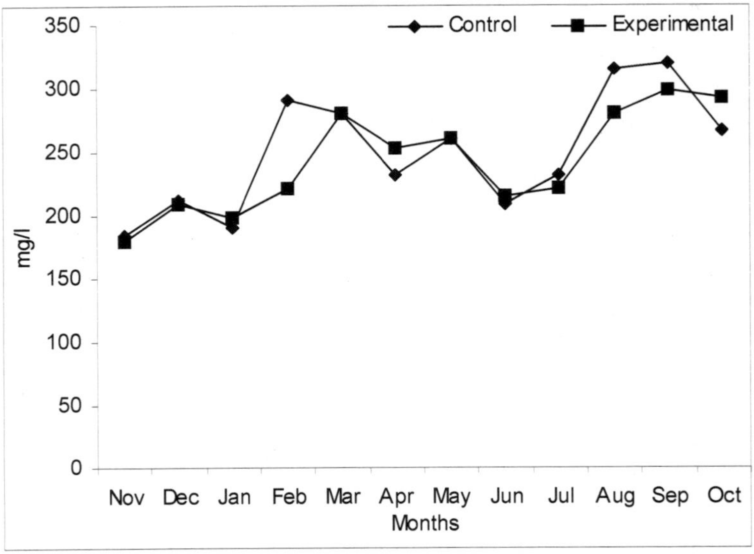

Figure 13: Seasonal Variation in Concentration of TDS

values in experimental (0.14±0.08 mg/l) and control pond (0.11±0.06 mg/l) are non-significantly different (Tables 7, 9 and Figure 14).

Comparatively higher value of NO_2-N were observed in control pond with the range of 0.018-0.068 mg/l, while it was in the range of 0.017-0.055 in experimental pond. One peak of value was observed during March-April in ponds. The pattern of fluctuation was similar in both the ponds. Mean values in experimental (0.04±0.01 mg/l) and control pond (0.05±0.02 mg/l) are significantly different (Tables 7, 9 and Figure 15).

Average highest values of NO_3-N concentration were observed in experimental pond. It was in the range of 0.07-1.22 and 0.1-0.94 mg/l in experimental and control pond, respectively. The concentrations of nitrate were low during winter season in both ponds. Mean values in experimental (0.78±0.32 mg/l) and control pond (0.57±0.26) are non-significantly different (Tables 7, 9 and Figure 16).

The phosphate concentrations in pond water ranged between 0.12-0.36 mg/l and 0.14-0.48 mg/l in experimental and control pond, respectively. On average values were higher in control pond with a

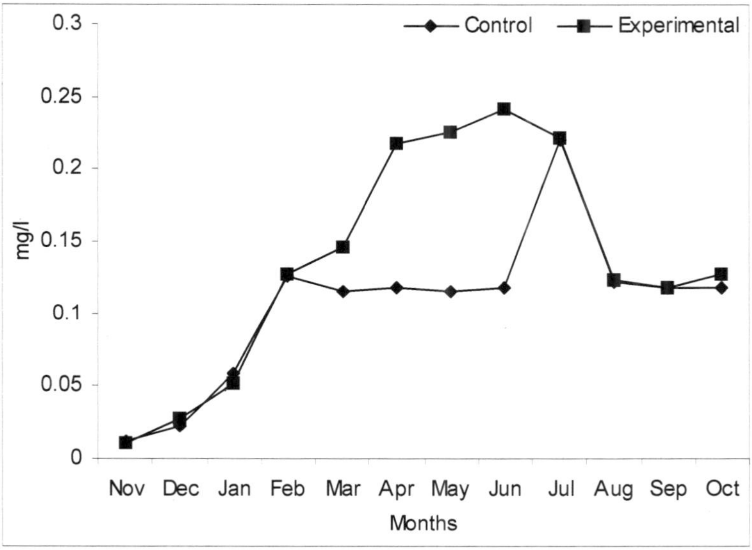

Figure 14: Seasonal Variation in Concentration of Ammonia Nitrogen

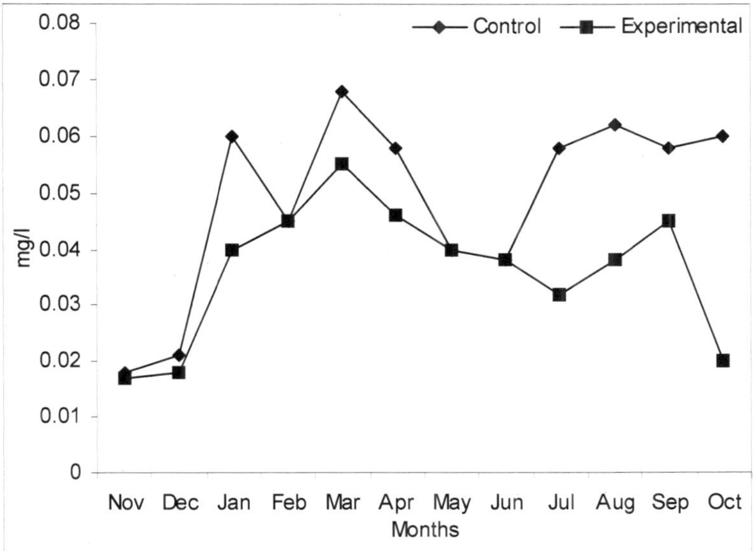

Figure 15: Seasonal Variation in Concentration of Nitrite Nitrogen

Table 10: Correlation Matrix of Different Physico-chemical Parameter of Experimental Pond Water and Gross Primary Productivity (GPP)

	Water Temp.	*pH*	*DO*	*CO_2*	*Total Alkalinity*	*Salinity*	*TDS*	*NH_4-N*	*NO_2-N*	*NO_3-N*	*Phos-phate*	*BOD_5*	*GPP*
Water temp.	1	-0.017	0.506	-0.430	0.402	0.549	0.468	0.816**	0.153	0.707*	0.040	-0.296	0.589*
pH		1	0.320	0.078	-0.680*	-0.455	-0.364	-0.342	-0.555	-0.394	-0.693*	0.490	-0.727**
DO			1	-0.114	0.527	0.047	0.742**	0.346	0.253	0.374	0.333	0.000	0.702*
CO_2				1	0.181	-0.552	0.244	-0.331	0.089	-0.137	0.038	0.704*	-0.325
Total alkalinity					1	0.557	0.606*	0.636*	0.666*	0.781**	0.614*	-0.207	0.738**
Salinity						1	0.053	0.775**	0.535	0.780**	0.536	-0.669*	0.651*
TDS							1	0.333	0.398	0.345	0.171	-0.071	0.563
NH_4-N								1	0.484	0.936**	0.392	-0.406	0.666*
NO_2-N									1	0.641*	0.721**	-0.238	0.435
NO_3-N										1	0.578*	-0.306	0.645*
Phosphate											1	-0.384	0.580*
BOD_5												1	-0.549

**: Significant at 1 per cent level; *: Significant at 5 per cent level.

Table 11: Correlation Matrix of Different Physico-chemical Characteristics of Control Pond Water and Gross Primary Productivity (GPP)

	Water Temp.	*pH*	*DO*	*CO_2*	*Total Alkalinity*	*Salinity*	*TDS*	*NH_4-N*	*NO_2-N*	*NO_3-N*	*Phos-phate*	*BOD_5*	*GPP*
Water temp.	1	0.654*	0.394	-0.432	0.255	0.547	0.299	0.638*	0.231	0.412	0.004	0.202	0.575
pH		1	-0.065	-.850**	-0.275	0.499	-0.252	0.046	-0.432	0.005	-0.183	-0.527	0.235
DO			1	-0.152	0.374	0.059	0.419	0.440	0.797**	0.534	0.479	0.545	0.673*
CO_2				1	0.362	-0.596*	0.528	0.073	0.498	0.120	0.097	0.709**	-0.161
Total alkalinity					1	0.407	0.391	0.374	0.454	0.640*	0.522	0.593*	0.664*
Salinity						1	-0.147	0.235	-0.070	0.588*	0.353	-0.118	0.596*
TDS							1	0.462	0.545	0.368	-0.108	0.797**	0.322
NH_4-N								1	0.630*	0.344	-0.051	0.676*	0.323
NO_2-N									1	0.476	0.286	0.827**	0.375
NO_3-N										1	0.690*	0.505	0.754**
Phosphate											1	0.189	0.671*
BOD_5												1	0.311

**: Significant at 1 per cent level; *: Significant at 5 per cent level.

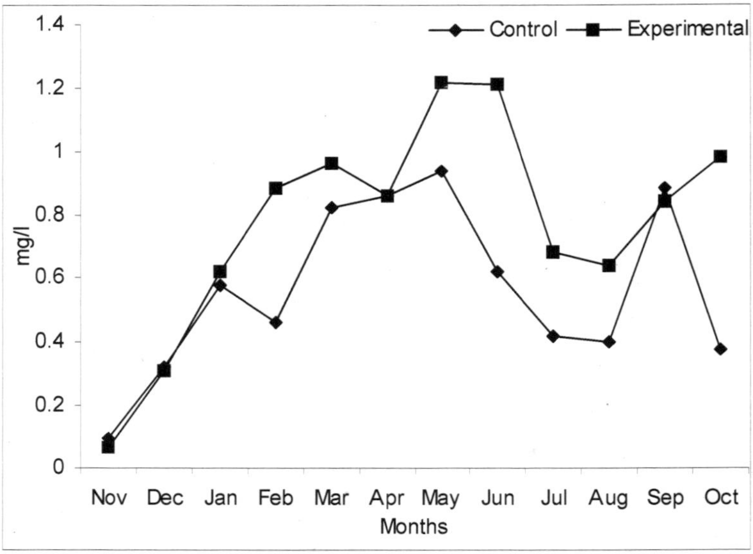

Figure 16: Seasonal Variation in Concentration of Nitrate Nitrogen

peak in summer months. Mean values in control (0.3±0.13 mg/l) and experimental pond (0.23±.08 mg/l) are significantly different (Tables 7, 9 and Figure 17).

Significant difference was observed in the values of BOD_5 between experimental and control pond. These values were highest with a range of 7.5-20.5 mg/l in control pond while this range was 5.0-11.5 mg/l in experimental pond. Highest values were observed during the months of February and September in both ponds. Mean values in experimental (6.15±3.64 mg/l) and control pond (16.58±3.85 mg/l) are significantly different (Tables 7, 9 and Figure 18).

A significant positive correlation with GPP was observed for water temperature, DO, total alkalinity, salinity, ammonia-nitrogen, nitrite, nitrate and phosphorus in both ponds. But, free CO_2 and BOD showed negative correlation with GPP.

Biotic Parameters in Pond Water

In view of diverse physical and chemical environment, the biotic community including plankton, benthos and primary productivity

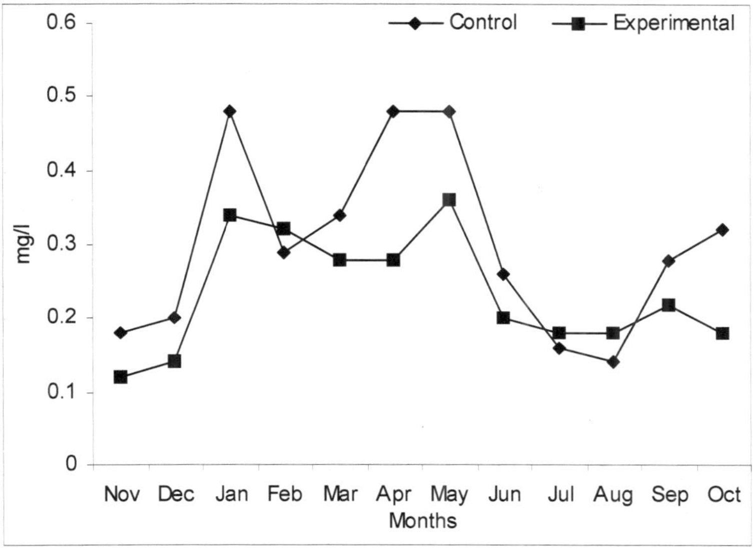

Figure 17: Seasonal Variation in Concentration of Phosphate

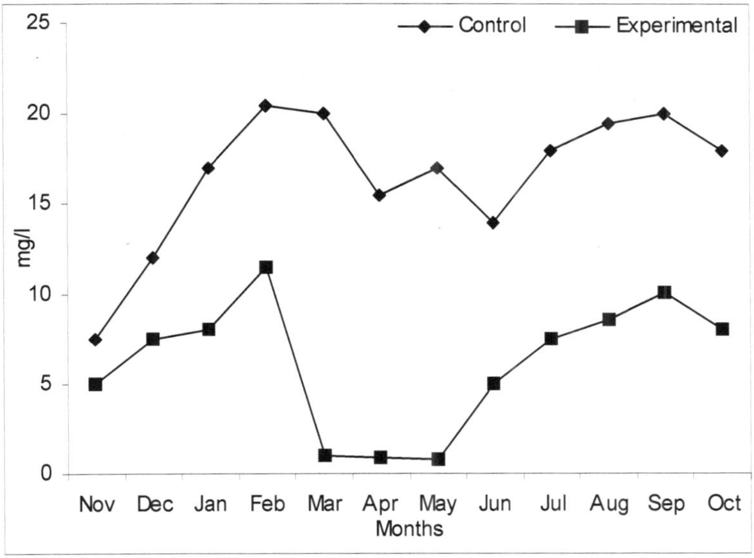

Figure 18: Seasonal Variation in Concentration of BOD_5

were analysed in both ponds. The observation recorded are presented in Table 12-19 and Figures 19-27, depict the seasonal variation of these parameters.

The averages value of phytoplankton and zooplankton are presented in Table 12-14, while Figure 22, exhibits their seasonal fluctuations. Table 17-19, depict the species composition and distribution pattern of plankton community in both ponds.

A total 18 genera of phytoplankton were recorded including 8 genera of chlorophyceae, 4 genera of bacillariophyceae, 3 genera of cynophyceae, 2 genera of euglenophyceae and one genera of dinophyceae. In experimental pond, maximum numbers was observed for green algae (60.3 per cent) followed by diatoms. While in control pond maximum number was observed for green algae (52.9) followed blue-green algae. The commonly ocuring green algae included *Pediastrum* spp., *Chlamydomonas* sp., *Volvox* sp, *Scenedesmus* spp. and *Chlorella* spp. All these species were observed thriving well in the ponds through out experimental period with their dominance in rainy and summer months. Diatoms were represented mainly by *Navicula* spp., *Nitzschia* spp., *Fragilaria* spp. *and Pinnularia* sp. Invariably, diatoms were most dominant in rainy season and their occurance become sparse in the winter months. Diatoms contributed 20.2 per cent in experimental in 14.8 per cent in control pond of total composition of phytoplankton. *Fragilaria* sp. was more prominent in experimental pond, while *Pinnularia* sp. was prominent in control pond. Among the blue green algae *Microcystis* sp. *Anabaena* sp. and *Nostoc* sp. were the major species in both ponds. Percent composition of blue green algae was comparatively higher (18.8 per cent) in control pond. Blue green algae were dominantly found in winter season and rarely occured in rainy and summer seasons in both ponds. They contributed 7.3 per cent in experimental and 9.9 per cent in control pond of the total density of phytoplankton. *Microcystis* sp., *Ananbaena* sp. and *Nostoc* sp. were major species of blue-green algae. *Anabaena* sp. was more prominent in experimental, while *Microcystis* sp. was prominent in control pond. Dinoflagellates were represented by *Peridinium* spp., which was more prominent in control pond and observed through out the year with higher values during the month of July-September. Among *Euglenofids viz.*, *Euglena* sp. and *Phacus* spp. were registered in both ponds. Among these, *Euglena* sp. was preponderant in both ponds. They contributed 7.3

Table 12: Seasonal Variation in Biotic Parameters of Experimental Pond Water

Month	Dipteran			Oligochaets			Molluscs No./m²	Plankton Vol. (ml/ 50 l.)	Phyto-plankton (nos./l.)	Zoo-plankton (nos./l.)	NPP gC/m³/ hr	CR gC/m³/ hr	GPP gC/m³/ hr
	No./ m²	Wet Wt. (mg/m²)	Dry wt. (mg/m²)	No./ m²	Wet Wt. (mg/m²)	Dry wt. (mg/m²)							
Nov	827	2026	260	65	1060	206	93	0.6	210	160	.115	.093	.208
Dec	1036	2560	370	64	1480	244	154	1.4	410	210	.116	.104	.220
Jan	1172	2890	410	136	3040	494	100	1.3	840	280	.185	.143	.328
Feb	1310	3163	468	200	4481	752	109	2.9	860	360	.284	.206	.480
Mar	1560	3736	566	201	4630	760	263	3.8	2000	580	.380	.240	.620
Apr	2030	5030	807	281	6290	744	218	4.2	3700	636	.710	.530	1.240
May	2645	6400	868	390	8664	1364	254	2.6	2600	580	.512	.508	1.020
Jun	2450	5898	812	350	8460	1298	218	2.4	1680	408	.318	.342	.660
Jul	2680	6580	910	181	4340	980	268	1.8	1420	402	.152	.168	.320
Aug	1810	3810	489	218	6290	686	236	2.1	1024	382	.202	.218	.420
Sep	2040	4822	788	198	4480	740	381	1.4	810	210	.301	.379	.680
Oct	2080	4826	789	222	4682	786	468	1.4	620	186	.402	.418	.820

Table 13: Seasonal Variation in Biotic Parameters of Control Pond Water

Month	*Dipteran*			*Oligochaets*			*Moll-uscs No./m²*	*Plankton Vol. (ml/ 50 l.)*	*Phyto-plankton (nos./l.)*	*Zoo-plankton (nos./l.)*	*NPP gC/m³/ hr*	*CR gC/m³/ hr*	*GPP gC/m³/ hr*
	No./ m²	*Wet Wt. (mg/m²)*	*Dry wt. (mg/m²)*	*No./ m²*	*Wet Wt. (mg/m²)*	*Dry wt. (mg/m²)*							
Nov	789	1886	256	48	1086	168	42	0.4	320	168	.115	.095	.210
Dec	988	2410	368	66	1490	262	136	1.2	400	220	.128	.098	.226
Jan	1210	2898	412	128	2840	398	148	1.3	480	180	.180	.140	.320
Feb	1302	3090	450	198	4262	668	144	2.8	680	220	.231	.199	.430
Mar	1480	3638	558	199	4425	710	108	3.2	1800	446	.330	.230	.560
Apr	2018	5010	806	283	6310	802	198	3.6	2600	620	.595	.455	1.050
May	2640	6408	882	271	8210	1102	180	2.8	2120	510	.465	.455	.910
Jun	2390	5680	808	298	7826	906	108	2.1	900	410	.260	.270	.530
Jul	2598	6466	902	172	4186	810	110	1.6	880	480	.145	.165	.310
Aug	1802	3806	4868	222	6310	712	110	1.8	920	320	.215	.245	.460
Sep	2020	4802	788	187	4188	608	126	1.2	620	240	.275	.315	.590
Oct	2098	4910	692	182	4082	678	124	1.6	480	180	.322	.358	.680

per cent in experimental in 9.9 per cent in control pond of the total density (Tables 12-14, 17-19 and Figure 22).

Table 14: Mean values (± SD) of different biotic parameters in experimental (E) and control (C) pond.

Parameters	*Pond*	
	E	*C*
Dipteran (Nos. / m²)	1803.33[a] ±626.85	1777.91[b] ± 621.53
Oligochaets (Nos. / m²)	208.83[a] ± 98.27	187.83[b] ± 78.49
Molluscs (Nos. / m²)	230.17[a] ± 112.40	127.83[a] ± 39.53
Plankton. (ml / 50 l)	2.16[a] ± 1.07	1.97[b] ± 0.95
Phytoplankton (Nos. / l)	1347.83[a] ± 1010.64	1016.67[b] ± 744.18
Zooplankton (Nos. / l)	366.17[a] ± 164.80	332.83[b] ± 154.45
GPP (gc / m³ / hr)	0.58[a] ± 0.30	0.52[b] ± 0.26
NPP (gc / m³ / hr)	0.31[a] ± 0.18	0.27[b] ± 0.15
CR (gc / m³ / hr)	0.28[a] ± 0.16	0.25[b] ± 0.12

Figures having same superscript in each row are significantly different ($P < 0.05$)

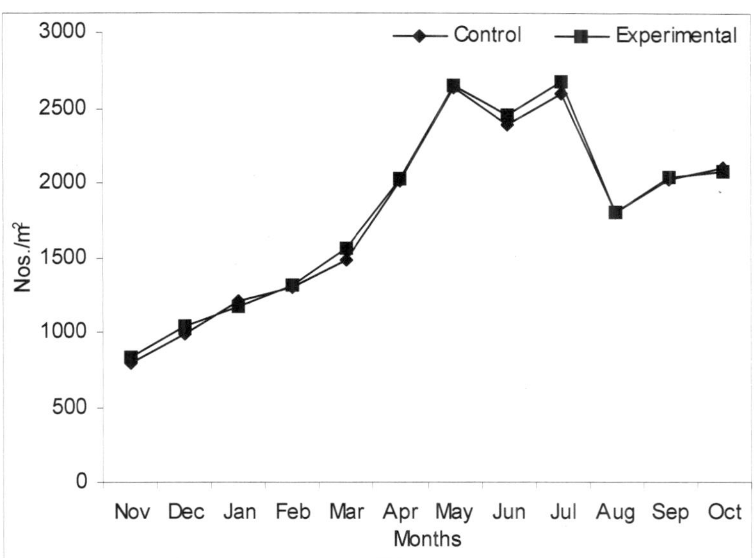

Figure 19: Seasonal Abundance in Population of Deptran

The total plankton volume varied from 0.6-4.2 ml/50 l. and 0.4-3.6 ml/50 l in experimental and control pond, respectively. Mean values in experimental (2.16±1.07 ml/50 litre) and control pond (1.97±0.95 ml/50 litre) are non-significantly different. Maximum value was observed during March-April in both ponds. Similar fluctuation pattern was observed in total phytoplankton count. Total density varied from 210-3700 nos./l and 320-2600 nos./ litre in experimental and control pond, respectively. Highest density were observed in the month of April with appearance of algal bloom (Plate 22) and minimum during November in both ponds. Mean values in experimental (1347.83±1010.64 nos/l) and control pond (1016.67±744.18 nos./l) are non-significantly different (Tables 12-14, Figure 24 and Plate 8).

A total of 7 species of zooplankton were collected during the course of the present study including 2 species of rotifera, 3 species of cladocera and 2 species of copepoda. The number of species present in different months varied through out the year. The maximum number of zooplankton was observed during winter

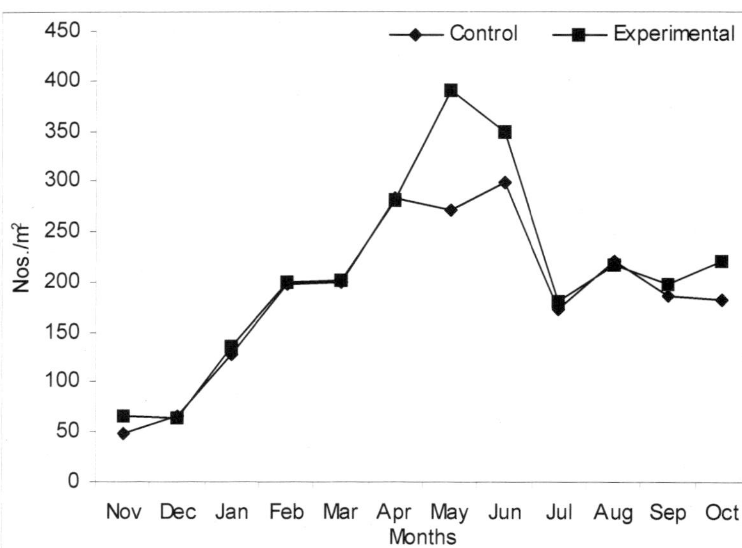

Figure 20: Seasonal Abundance in Population of Oligochaete

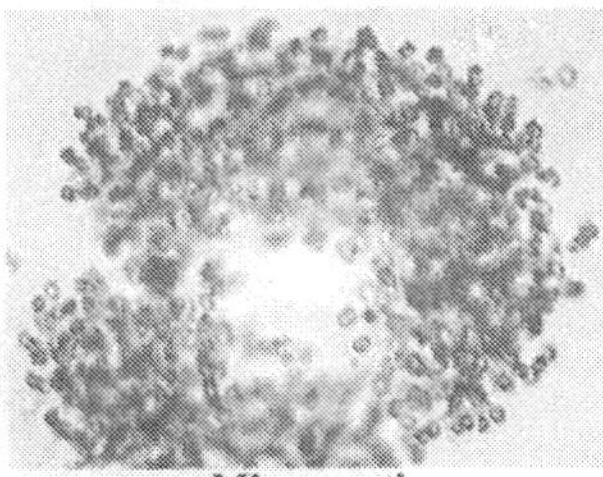

Microcystis

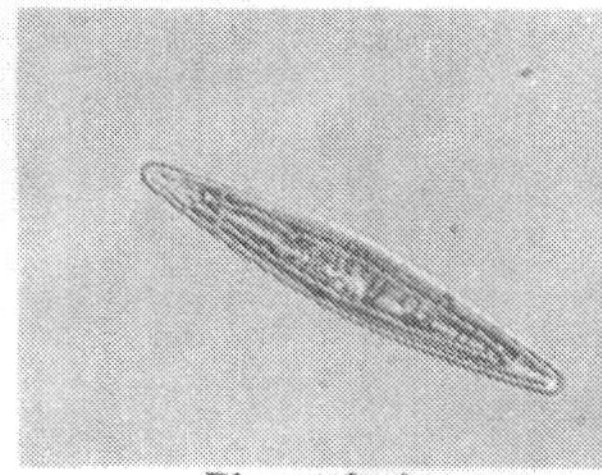

Pinnularia

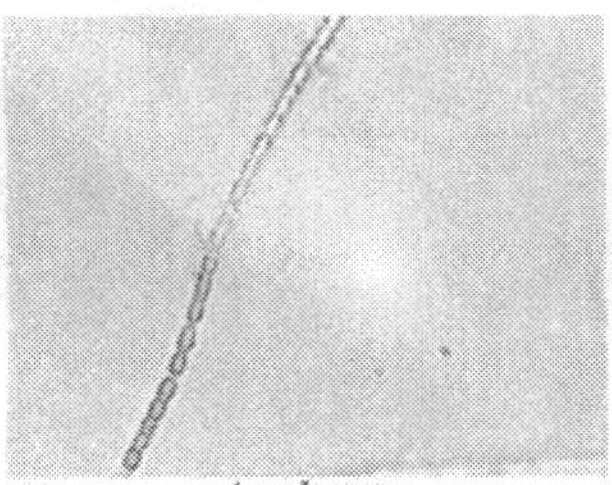

Anabaena

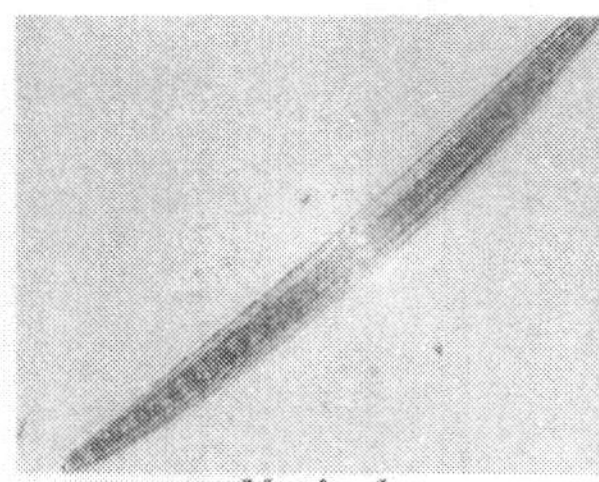

Navicula

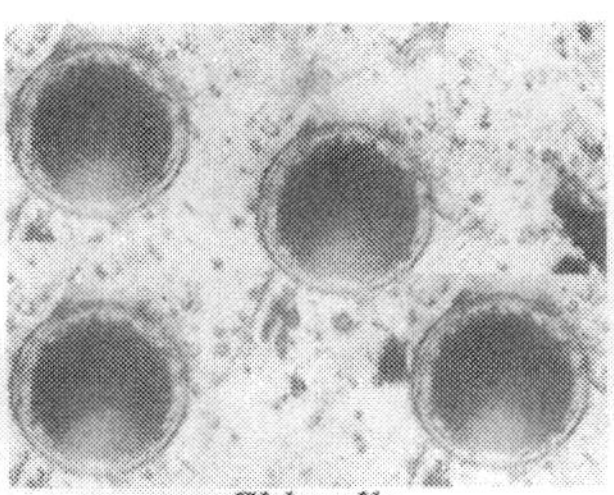

Chlorella

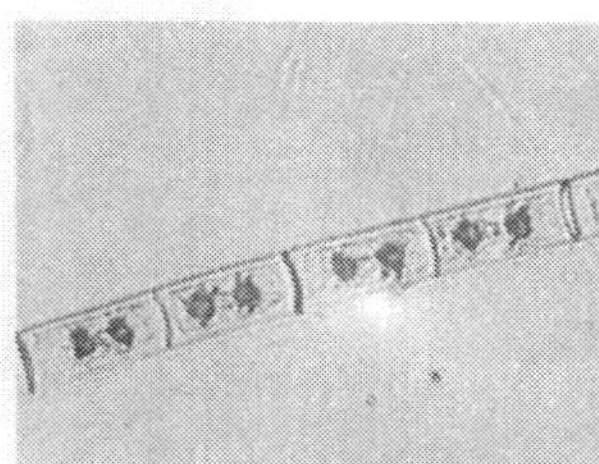

Zygnema

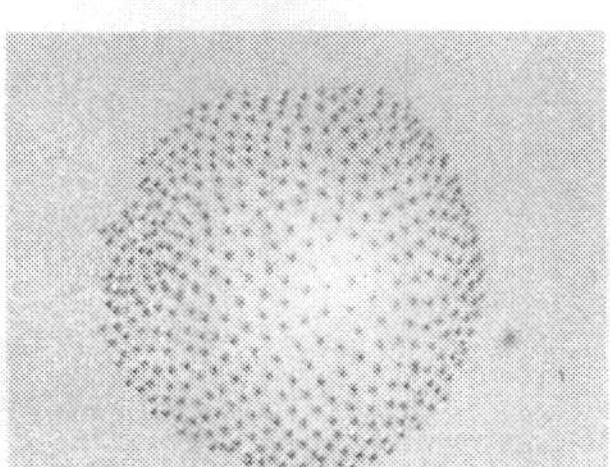

Volvox

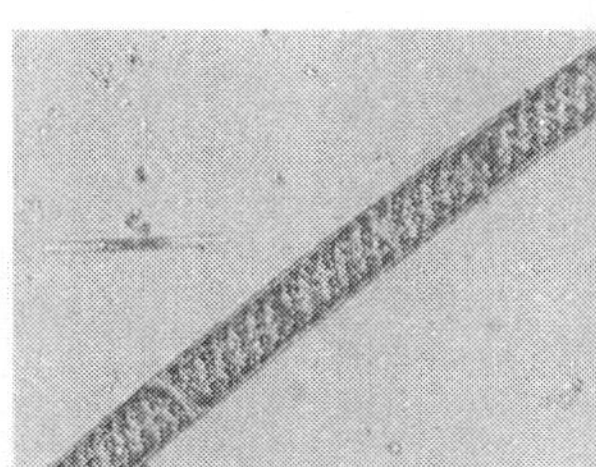

Spirogyra

Plate 8: Observed Phytoplankton in Experimental Pond

Table 15: Correlation Matrix of Biotic Parameters in Experimental Pond and Gross Primary Productivity (GPP)

	Dipteran Density	*Oligochaets Density*	*Molluscs density*	*Plankton Vol.*	*Phyto-plankton Density*	*Zooplankton Density*	*NPP*	*CR*	*GPP*
Dipteran density	1	0.798**	0.607*	0.293	0.541	0.475	0.463	0.668*	0.573
Oligochaets density		1	0.356	0.557	0.705*	0.671*	0.714**	0.818**	0.779*
Molluscs density			1	0.016	0.094	0.022	0.353	0.580*	0.471
Plankton vol.				1	0.839**	0.902**	0.749**	0.507	0.649*
Phytoplankton density					1	0.926**	0.858**	0.714**	0.810**
Zooplankton density						1	0.704*	0.519	0.632*
NPP							1	0.911**	0.980**
CR								1	0.975**

**: Significant at 1 per cent level; *: Significant at 5 per cent level.

Table 16: Correlation Matrix of Biotic Parameters in Control Pond and Gross Primary Productivity (GPP)

	Dipteran Density	*Oligochaets Density*	*Molluscs density*	*Plankton Vol.*	*Phyto-plankton Density*	*Zooplankton Density*	*NPP*	*CR*	*GPP*
Dipteran density	1	0.760**	0.363	0.362	0.449	0.641*	0.485	0.688*	0.589*
Oligochaets density		1	0.535	0.751**	0.688*	0.711**	0.737**	0.795**	0.779**
Molluscs density			1	0.656*	0.621*	0.481	0.723**	0.651*	0.702**
Plankton vol.				1	0.855**	0.739**	0.791**	0.628*	0.729**
Phytoplankton density					1	0.897**	0.882**	0.727**	0.825**
Zooplankton density						1	0.693*	0.591*	0.658*
NPP							1	0.924**	0.984**
CR								1	0.977**

**: Significant at 1 per cent level; *: Significant at 5 per cent level.

season followed by summer. The percent composition of rotifers and cladocerans was almost equal (40-41.5 per cent) in experimental pond but it was quite different in control pond which was dominated (61 per cent) by rotifers. Percent composition of *Cladoceran* was higher (40.1 per cent) in experimental pond than in control (20.3 per cent) pond. Minimum composition was observed for *Copepods* in both ponds. Maximum number of rotifers was noted in December, *Cladocerans* were present throughout the year with peak in winter season. Two species of rotifers namely *Brachionus* spp.and *Keratella* spp. consistently occurred in zooplankton population. *Cladoceran* population included individuals of *Daphnia* spp. *Moina* sp. and *Bosmina* sp. Among them *Daphnia* sp. and *Moina* sp. were noticed round the year in both ponds. *Bosmina* sp. was predominant during winter and summer. Copepod population consisted of *Cyclops* spp. and *Diaptomus* sp. with abundance in winter and summer (Tables 12-14, 17-19 and Figure 23).

The density of zooplankton progressively with the start of the experiment in both ponds ranging on peak in January and April. The density varied from 160-636 nos./l and 168-620 nos./l in

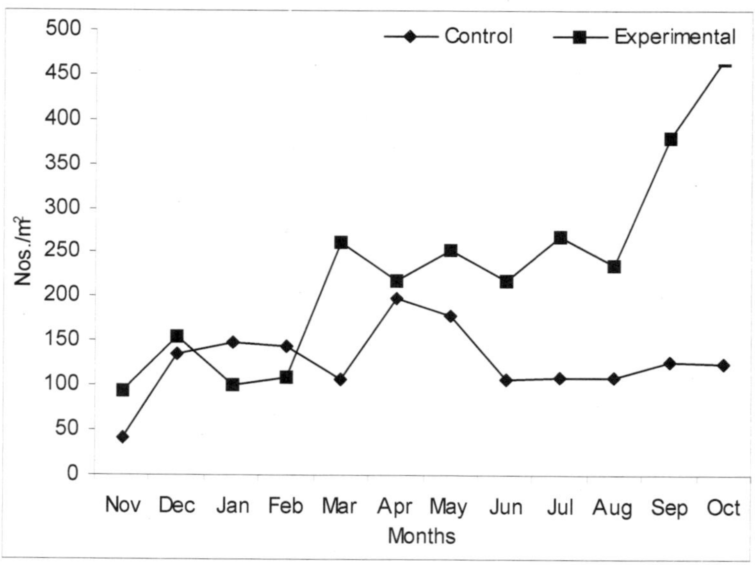

Figure 21: Seasonal Abundance in Population of Molluscs

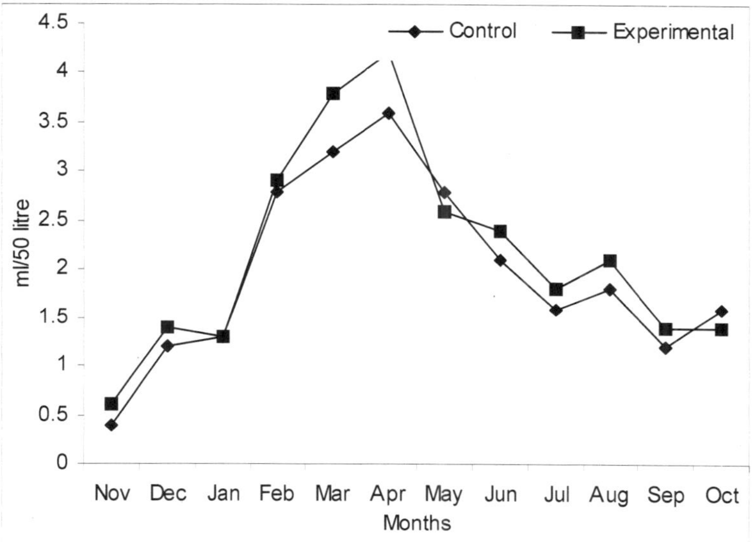

Figure 22: Seasonal Variation in Plankton Volume

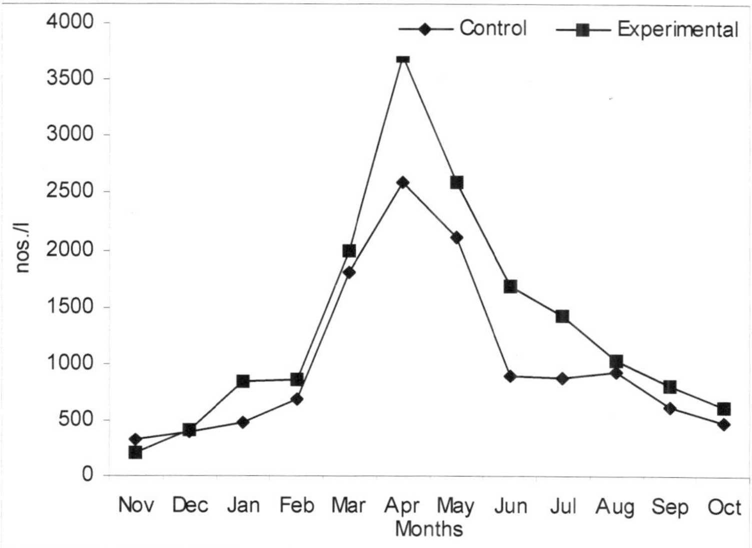

Figure 23: Seasonal Variation in Phytoplankton Population

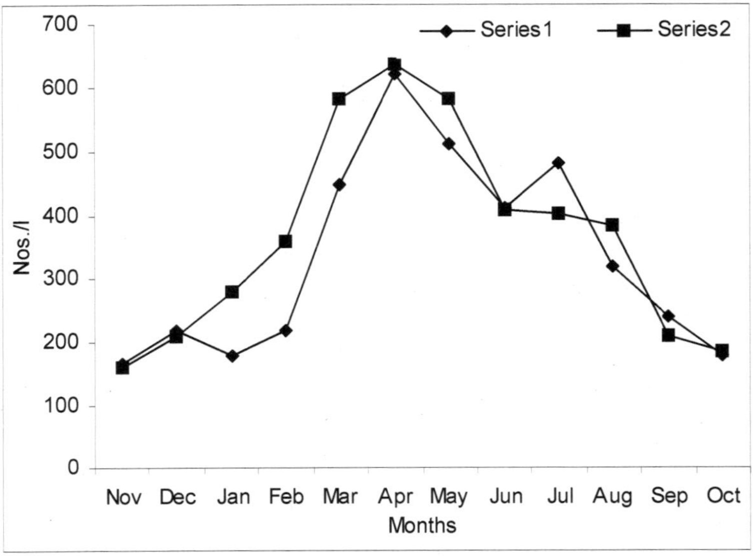

Figure 24: Seasonal Variation in Zooplankton Population

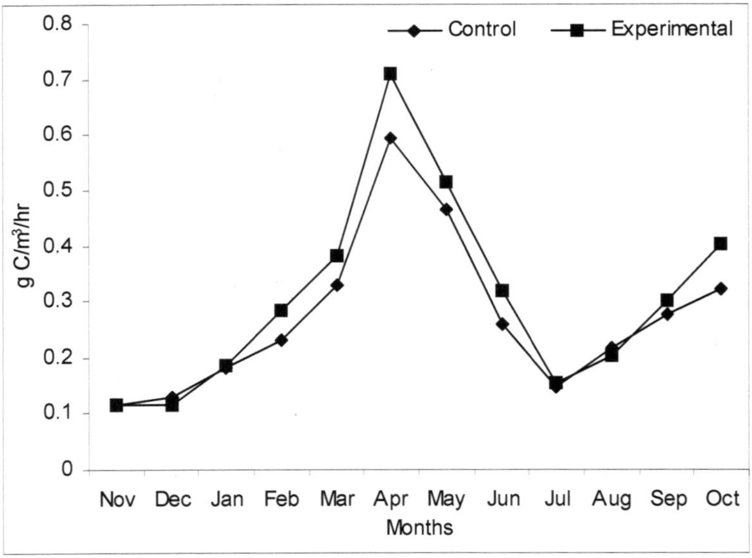

Figure 25: Seasonal Variation in NPP Level

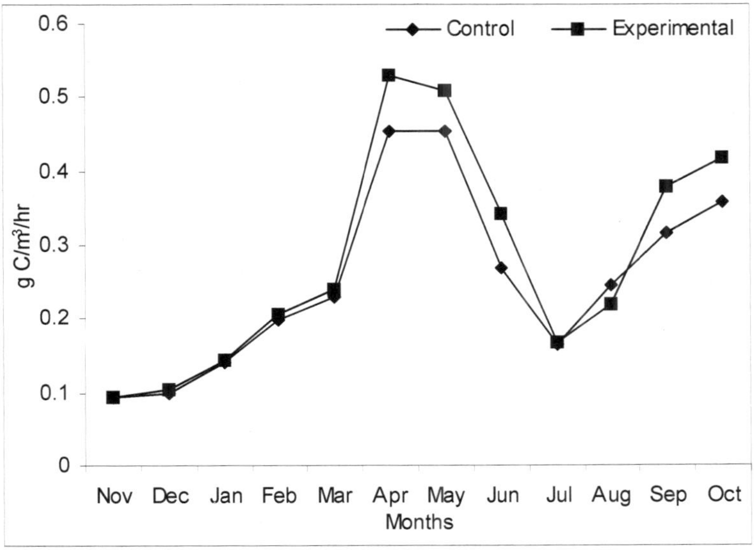

Figure 26: Seasonal Variation in CR

experimental and control pond, respectively. The total density of zooplankton was comparatively higher in experimental pond. Mean values in experimental (366.17±164.8 nos./l) and control pond (332.83±154.45 nos./l) are non-significantly different (Table 12-14, Figure 23 and Plate 9).

Results pertaining to numerical density, biomass and seasonal fluctuation of benthic fauna are presented in Table 12-14 and Figure 19-21. Benthic fauna recorded in present experiment was comprised of mollusca, insecta and oligochaeta. Diptera represented by *Chironomus* sp. with varied range of 827-2680 (1803.33±626.85) nos./ m^2 and 789-2640 (1777.91±621.53) nos./m^2 in experimental and control pond, respectively. A gradual increasing trends was observed from starting of experiment in total density of deptera, which was again decreased after the month of July. Mean values in experimental.and control pond are non-significantly different (Figure 19 and Plate 10).

Oligochaete were represented by *Tubifex* sp. and *Dero* sp. with varied range of 65-390 (208.83±98.27) nos./m^2 and 48-371 (187.83±78.49) nos./m^2 in experimental and control pond.

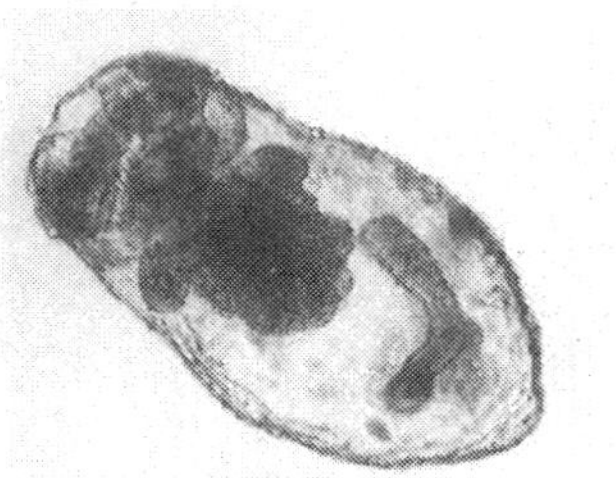

Asplanchana

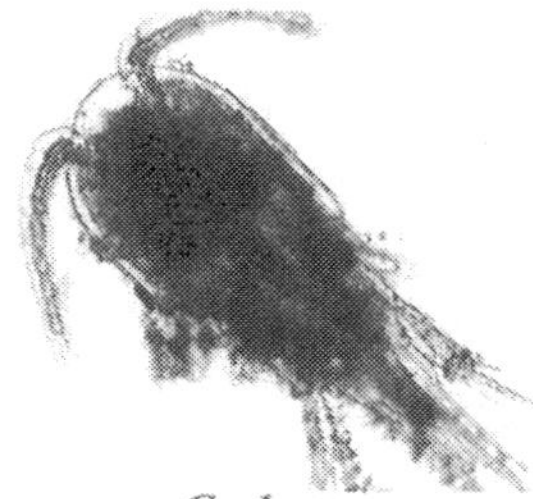

Cyclops

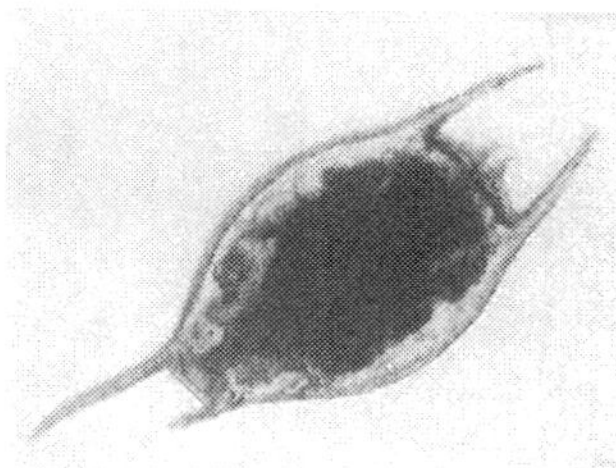

Keratella

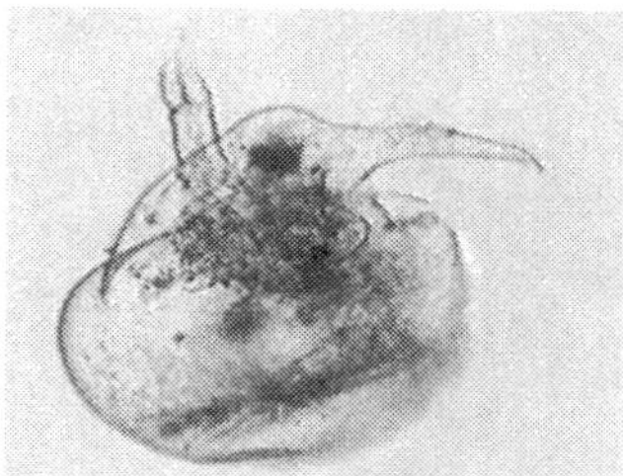

Bosmina

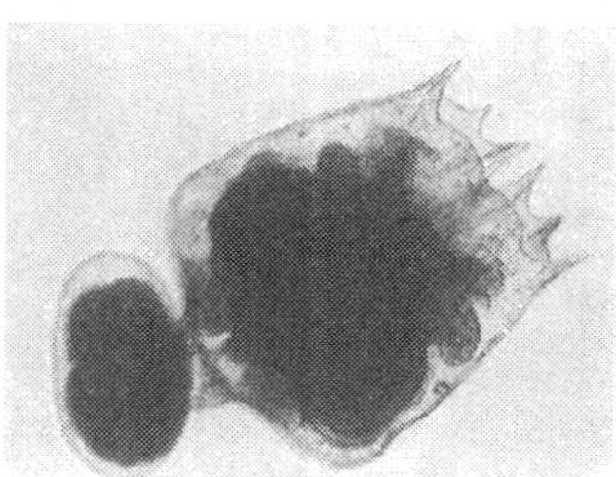

Branchionus

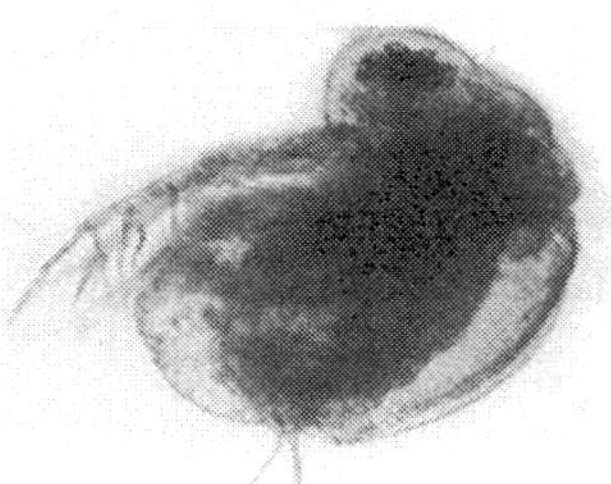

Moina

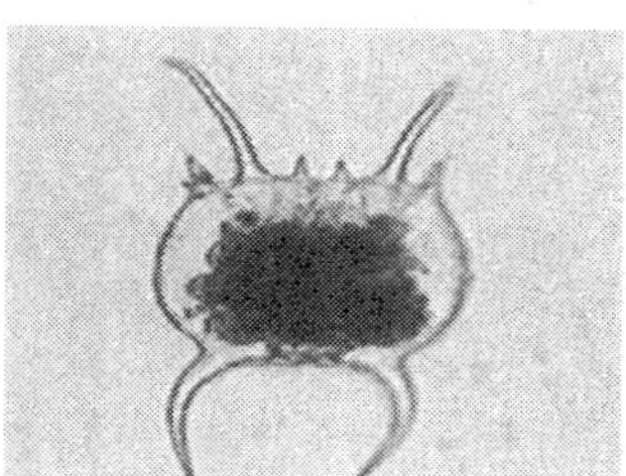

Branchionus

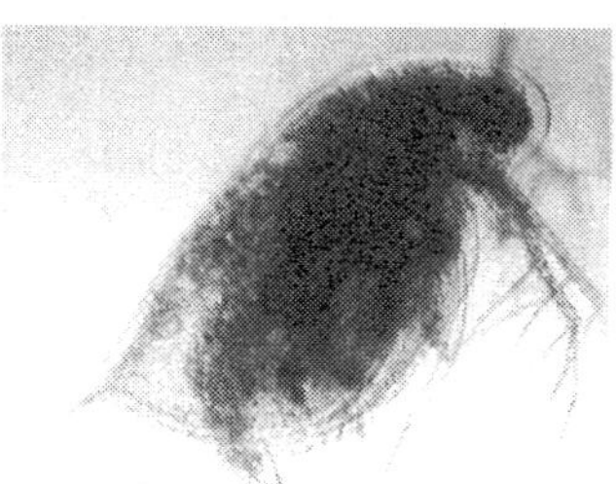

Daphnia

Plate 9: Observed Zooplankton in Experimental Pond

Plate 10: Observed Benthic Fauna in Experimental Pond

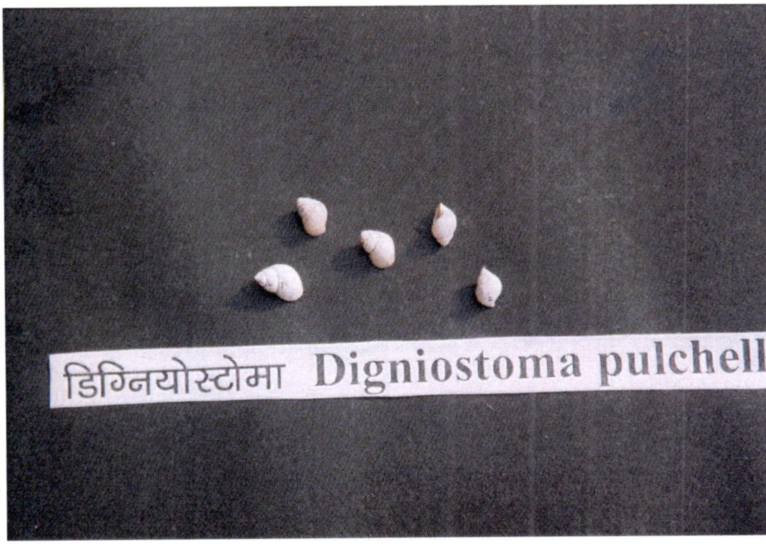

Contd...

Plate 10–Contd...

Contd...

Plate 10–Contd...

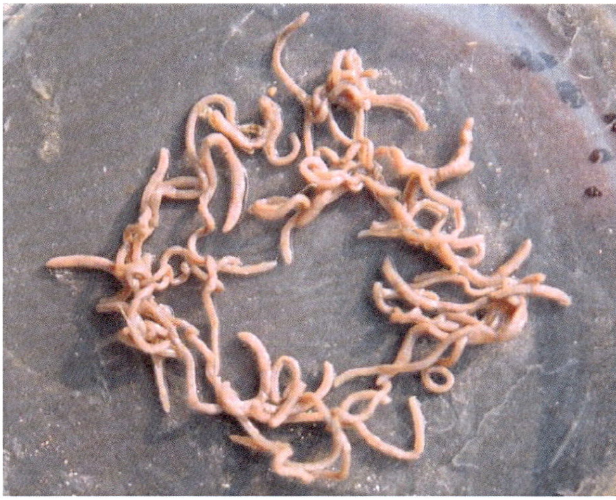

Maximum density of oligochaets was observed during the month of May. Common gastropods found in both ponds were *Gyraulus* sp. *Limnaea* sp. *Digonostoms* sp. and *Vivipara* sp. only one species (*Lamellidens* sp.) was noticed under the pelecypods. Mean values in experimental and control pond are non-significantly different. Total density of molluscs varied from 93-263 (230.17±112.40) nos./m^2 and 42-208 (127.83±39.53) nos./m^2 in experimental and control pond, respectively. Mean values in experimental and control pond are significantly different. Overall density of benthic fauna is higher in experimental pond in comparison to control pond.

On biochemical analysis, plankton showed 68 per cent moisture and 55 per cent protein and banthose showed 85 per cent moisture and 50.1 per cent protein on dry matter base.

GPP showed wide seasonal fluctuation ranging between 0.208-1.240 (0.58±0.30) gC/m^3/hr in experimental pond and 0.210-1.050 (0.52±0.26) gC/m^3/hr in control pond. It increased rapidly from the beginning of the experiment reaching peak values in the months of April-May in both ponds. From August onwards again, the values of GPP started increasing attaining second peak in the month of October. Intensity of the first peak was higher than the second peak

Table 17: Species Composition and Distribution Pattern of Plankton in Experimental Pond

Species	*Months*											
	Nov	*Dec*	*Jan*	*Feb*	*Mar*	*Apr*	*May*	*Jun*	*Jul*	*Aug*	*Sep*	*Oct*
Phytoplankton												
(A) Chlorophyceae												
Pediastrum spp.					+	+		+	+			
Chlamydmonas sp.	+	+	+	+		+	+	+	+	+	+	+
Volvox sp.			+				+		+	+		
Scenedesmus spp.	+				+	+	+	+		+	+	
Chlorella spp.	+			+	+	+	+		+	+	+	+
Spirogyra sp.		+	+	+					+			
Oedogonium sp.		+	+	+								
Cosmarium spp.		+				+	+		+			
(B) Bacillariophyceae												
Navicula spp.	+			+		+	+	+	+	+	+	
Nitzschia spp.					+	+			+		+	
Fragilaria spp.			+				+	+	+	+		+
Pinnularia sp.		+							+	+	+	

Contd...

Table 17–Contd...

Species	*Months*											
	Nov	*Dec*	*Jan*	*Feb*	*Mar*	*Apr*	*May*	*Jun*	*Jul*	*Aug*	*Sep*	*Oct*
(C) Cynophyceae												
Microcistis sp.		+	+	+	+			+			+	
Anabaena sp.	+	+	+	+		+				+		+
Nostoc sp.		+	+	+				+				
(D) Euglenophyceae												
Euglena sp.	+	+	+	+	+	+					+	+
Phacus sp.		+	+			+		+				
(E) Dinophyceae												
Peridinium spp.				+	+			+	+	+	+	
Zooplankton												
(A) Rotifera												
Brachionus spp.	+	+	+		+	+		+	+		+	+
Keratella spp.		+	+			+	+					
(B) Cladocera												
Daphnia spp.	+	+	+	+	+	+	+		+	+	+	
Moina sp.	+	+	+	+	+	+	+		+	+	+	
Bosmina sp.		+	+			+						
(C) Copepoda												
Cyclops spp.		+	+	+		+	+	+				
Diaptomus sp.		+	+			+			+			

Table 18: Species Composition and Distribution Pattern of Plankton in Control Pond

Species	*Months*											
	Nov	*Dec*	*Jan*	*Feb*	*Mar*	*Apr*	*May*	*Jun*	*Jul*	*Aug*	*Sep*	*Oct*
Phytoplankton												
(A) Chlorophyceae												
Pediastrum spp.					+	+		+	+			
Chlamydmonas sp.	+	+		+	+	+		+	+	+	+	+
Volvox sp.		+	+		+	+	+		+	+	+	
Scenedesmus spp.	+				+	+	+	+		+	+	
Chlorella spp.	+			+	+	+	+		+	+	+	+
Spirogyra sp.		+	+	+	+	+	+		+			
Oedogonium sp.		+	+	+			+					
Cosmarium spp.		+	+	+		+						
(B) Bacillariophyceae												
Navicula spp.	+			+		+	+	+	+	+	+	
Nitzschia spp.					+						+	
Fragilaria spp.								+	+	+		+
Pinnularia sp.		+				+		+	+		+	

Contd...

Table 18–Contd...

Species	*Months*											
	Nov	*Dec*	*Jan*	*Feb*	*Mar*	*Apr*	*May*	*Jun*	*Jul*	*Aug*	*Sep*	*Oct*
(C) Cynophyceae												
Microcistis sp.		+	+	+	+			+			+	
Anabaena sp.	+	+				+		+		+		+
Nostoc sp.		+	+		+					+		
(D) Euglenophyceae												
Euglena sp.	+	+	+		+	+	+	+		+	+	+
Phacus sp.		+	+			+	+	+				
(E) Dinophyceae												
Peridinium spp.			+	+	+		+	+	+	+	+	
Zooplankton												
(A) Rotifera												
Brachionus spp.	+	+	+	+	+	+	+	+	+	+	+	+
Keratella spp.	+	+	+			+	+					
(B) Cladocera												
Daphnia spp.	+	+	+		+	+	+		+		+	
Moina sp.	+	+		+	+		+		+	+	+	
Bosmina sp.		+	+			+						
(C) Copepoda												
Cyclops spp.		+		+		+	+	+				
Diaptomus sp.		+	+		+	+			+			

in both pond. All biotic parameters have positive correlation with GPP in both ponds.

The mean values of NPP were obtain as 0.31±0.18 gC/m^3/hr and 0.27±0.15 gC/m^3/hr in experimental and control pond, respectively. It was minimum (0.115 gC/m^3/hr) in beginning and maximum (0.710 gC/m^3/hr) during the month of April in experimental pond. These values are non-significantly different. The perusal of seasonal variation of NPP revealed that period of the higher GPP values (April-May and September-October) coincided with the highest NPP value in both ponds. The intensity of first peak (April) was more prominent than the second peak (October) in both ponds. NPP is positively correlated with all biotic parameters with highly significant correlation with phytoplankton density at 1 per cent level. Community respiration ranged between 0.093-0.530 gC/m^3/hr and 0.095-0.455 gC/m^3/hr with mean value as 0.28±0.16 gC/m^3/hr and 0.25±0.12 gC/m^3/hr in experimental and control pond, respectively. Mean values are non-significantly different. Community respiration values were very less (0.093 gC/m^3/hr in experimental and 0.095 gC/m^3/hr in control pond) initially end increased gradually. The values were relatively low during winter months in both ponds. Mean values are non-significantly different ($P<0.05$) and it shows positive correlation with all biotic parameters

Table 19: Composition (%) of Plankton in Experimental and Control Pond

Plankton Group	*Experimental Pond*	*Control Pond*
Phytoplankton		
Chlorophyceae	60.3	52.9
Bacillarophyceae	20.2	14.8
Euglanophyceae	7.3	9.9
Cynophyceae	10.2	18.8
Dinophyceae	2.0	3.6
Zooplankton		
Rotifera	41.5	61.0
Cladocera	40.1	20.3
Copepoda	18.4	18.7

Fish Production

Observation on survival, growth and production of fish in pig-fish pond (control) and azolla-pig-fish (experimental) were taken for 12 months (November 2002-October 2003) of rearing period (Tables 20-27 and Figures 28-52).

The average survival rate of fishes in experimental pond was 86.63 per cent, while it was 83.27 per cent in control pond. *Cyprinus carpio* showed highest survival (98.95 per cent) followed by *Labeo rohita* (84.03 per cent) and *Cirrhinus mrigala* (82.29 per cent). Low survival was recorded for *Catla catla* (81.25 per cent and 78.65 per cent) in both ponds (Table 22)

The highest average final weight was attained by *Cyprinus carpio* in both ponds. Average net weight of this fish was 1045.5 g in experimental pond and 878.4 g in control pond. The average net weight of *Catla catla* was 844.2 g and 726.9 g experimental and control pond, respectively. *Labeo rohita* showed lower average net weight as 692.9 g and 576.6 g in experimental and control pond, respectively. The average net weight of *Cirrhinus mrigala* was 680.9 g in experimental and 564.4 g in control pond. Individual fish of highest weight was recorded as *Catla catla* (1350 g) in experimental, *Labeo rohita* (1200 g) in experimental, *Cirrhinus mrigala* (1040 g) in control and *Cyprinus carpio* (1600 g) in experimental pond (Table 22 and Plates 11-13).

The growth rate was comparatively less from October to January and increasing from March onwards (Figure 28-32). Species wise growth are presented in (Table 20,21). Specific growth rate (SGR) was highest for *Cirrhinus mrigala* in both ponds. The overall SGR showed in the range of 0.6246-0.7611 and 0.5483-0.7356 in experimental and control pond, respectively (Table 23).

The total fish production during 12 months was recorded as 637.8 kg. in experimental and 531.2 kg in control pond. The gross fish production was calculated to be 5315 kg./ha/yr. in experimental and 4427 kg/ha/yr. in control pond. In experimental pond, the total production was contributed as 40.53 per cent by *Catla catla*, 26.54 per cent by *Labeo rohita*, 17 per cent by *Cirrhinus mrigala* and 15.93 per cent by *Cyprinus carpio*. In control pond, it was 42.62 per cent by *Catla catla*, 25.83 per cent by *Labeo rohita*, 17.36 per cent by *Cirrhinus*

Plate 11: Experimental Pond Grown *Catla catla*

Plate 12: Experimental Pond Grown *Labeo rohita*

Table 20: Growth Performance of Fish in Experimental Pond

Months	*Catla catla*		*Labeo rohita*		*Cirrhinus mrigala*		*Cyprinus carpio*		*Average of All spp.*	
	*L**	*W***	*L*	*W*	*L*	*W*	*L*	*W*	*L*	*W*
Nov	15.7	74.0	16.5	69.1	14.9	45.1	19.1	119.7	16.6	77.0
Dec	18.2	95.0	18.9	98.5	15.8	58.0	23.6	193.0	19.1	111.1
Jan	19.9	123.0	22.0	139.5	17.6	84.0	26.0	343.0	21.4	172.4
Feb	23.9	195.5	24.7	179.0	19.4	104.0	27.0	400.0	23.8	219.6
Mar	26.9	281.0	26.2	211.5	23.8	228.0	28.0	440.0	26.2	290.1
Apr	28.1	319.0	27.3	242.0	28.2	304.0	29.8	490.0	28.4	338.8
May	28.9	372.0	28.7	328.0	29.4	330.0	31.0	566.0	29.5	399.0
Jun	31.6	495.0	30.5	396.0	32.0	384.0	36.6	780.0	32.7	513.8
Jul	35.0	588.0	34.3	530.0	33.6	468.0	38.4	920.0	35.3	626.5
Aug	37.8	760.0	36.8	640.0	35.8	508.0	39.0	1000.0	37.4	727.0
Sep	39.2	834.0	37.3	685.0	37.8	610.0	39.0	1012.0	38.3	785.3
Oct	39.6	918.2	39.1	762.0	39.8	726.0	41.0	1165.2	39.9	892.8

*: Length in cm; **: Weight in gm.

Table 21: Growth Performance of Fish in Control Pond

Months	*Catla catla*		*Labeo rohita*		*Cirrhinus mrigala*		*Cyprinus carpio*		*Average of All spp.*	
	*L**	*W***	*L*	*W*	*L*	*W*	*L*	*W*	*L*	*W*
Nov	15.9	74.5	16.6	70.1	14.4	41.3	20.6	137.2	14.4	80.8
Dec	17.7	90.5	19.1	98.0	15.8	57.0	21.2	154.0	18.5	99.9
Jan	19.4	103.0	21.2	126.5	17.4	82.0	25.0	300.0	20.8	152.9
Feb	23.4	154.0	23.3	149.0	19.2	102.0	26.0	320.0	23.0	181.3
Mar	26.1	266.0	25.8	198.0	22.4	161.0	28.0	405.0	25.6	257.5
Apr	27.0	288.0	26.5	222.5	28.0	257.0	28.6	460.0	27.5	306.9
May	28.6	324.0	27.4	265.0	28.0	274.0	30.4	508.0	28.6	342.8
Jun	30.5	415.5	30.3	365.0	31.8	370.0	36.0	710.0	32.2	465.1
Jul	32.0	526.0	31.8	435.0	33.2	460.0	36.8	780.0	33.4	550.3
Aug	36.5	685.0	35.7	569.0	35.8	486.0	38.6	920.0	36.7	665.0
Sep	38.0	735.0	36.6	600.0	36.6	538.0	38.8	930.0	37.5	700.8
Oct	38.5	801.4	37.6	646.7	38.5	605.7	39.8	1015.6	38.6	767.4

*: Length in cm; **: Weight in gm.

Plate 13: Fish Haul from Experimental Pond

mrigala and 14.19 per cent by *Cyprinus carpio*. At the end of experiment, 20.06 per cent increase was recorded in fish production over the control pond, in which 4 per cent increased due to better survival and 16.06 per cent increase was due to better productivity of pond. Maximum specie-wise increment was recorded for *Cyprinus carpio* (34 per cent), while it was lowest for *Catla catla* (14.2 per cent) (Table 22 and Figure 53).

The length-weight relationship was computed from the data collected thought out the investigation period of 12 months. The equations obtained are shown in (Tables 24-25). The scatter diagram obtained by plotting total lengths against weights of each species resulting in an experimental curve and the data for which the line of best fit was drawn by least square method are presented in Figures 33-48. The same for all fishes pooled in the pond is also depicted in Figures 49-52. Species wise correlation coefficient was calculated as 0.9727 for *Catla catla*, 0.9728 for *Labeo rohita*, 0.9834 for *Cirrhinus mrigala*, 0.9877 for *Cyprinus carpio* and 0.9836 for all fishes pooled in experimental pond. While, it was 0.9743 for *Catla catla*, 0.9771 for

Table 22: Survival, Growth and Production of Fish in Experimental (E) and Control (C) Ponds (Period of rearing 12 months, pond area 0.12 ha.)

Fish species	*Nos. Stocked*		*Nos. Harvested*		*Rate of Survival*		*Av. Initial Wt. (gm)*		*Av. Final Wt. (gm)*		*Growth Increment (gm)*		*Production*	
	C	*E*	*C*	*E*	*C*	*E*	*C*	*E*	*C*	*E*	*C*	*E*	*C*	*E*
Catla catla	384	384	302	312	78.65	81.25	74.5	74.0	801.4	918.2	726.9	844.2	226.4 (219.5)*	258.5 (263.4)*
Labeo rohita	288	288	230	242	79.86	84.03	70.1	69.1	646.7	762.0	576.6	692.9	137.2 (132.6)*	169.3 (167.7)*
Cirrhinus mrigala	192	192	159	158	80.81	82.29	41.26	45.13	605.67	726.0	564.4	680.9	92.2 (89.8)*	108.4 (107.6)*
Cyprinus carpio	96	96	90	95	93.75	98.95	137.17	119.67	1015.6	1165.2	878.4	1045.5	75.4 (75.0)*	101.6 (99.3)*

Total production (kg) E: 637.8; C: 531.2, Calculated production per ha. E: 5315; C: 4427, *Calculated quantity.

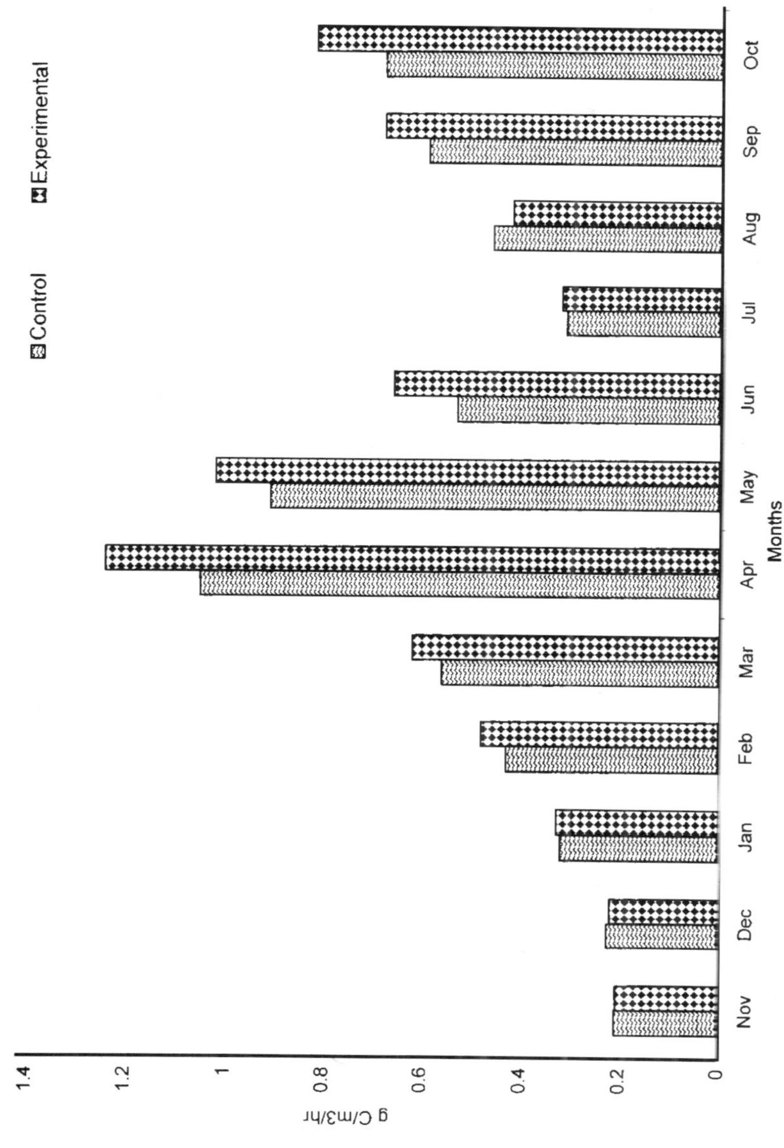

Figure 27: Seasonal Variation in GPP Level

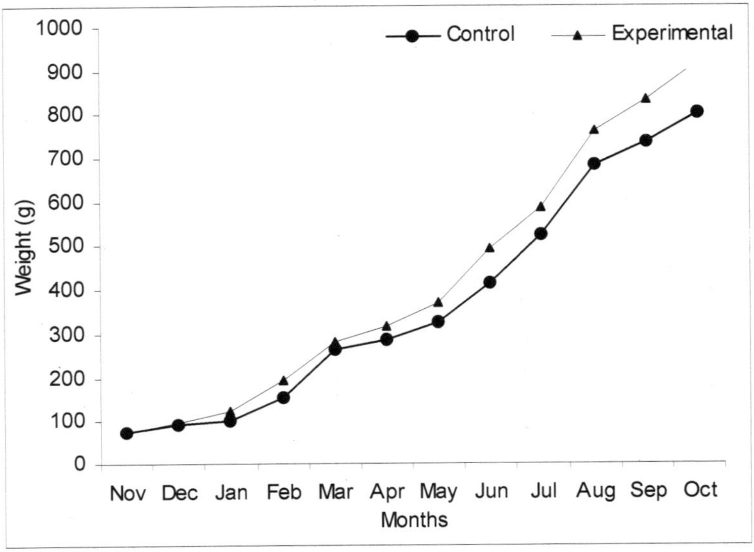

Figure 28: Growth of *Catla catla*

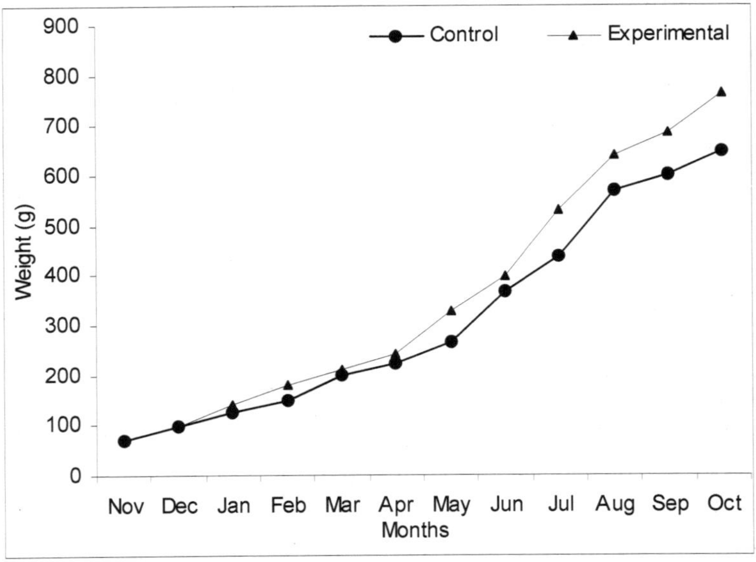

Figure 29: Growth of *Labeo rohita*

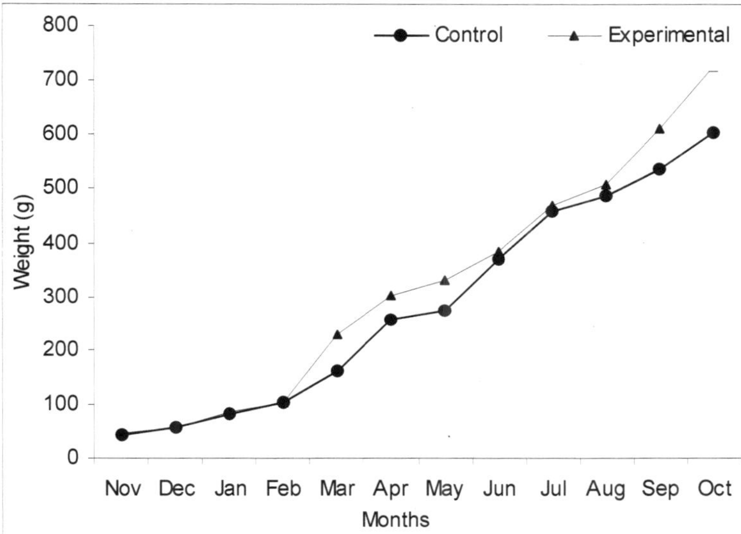

Figure 30: Growth of *Cirrhinus mrigala*

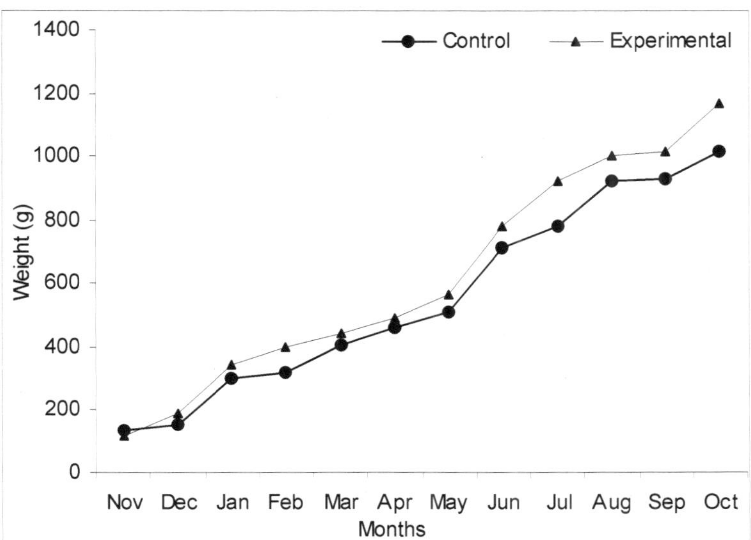

Figure 31: Growth of *Cyprinus carpio*

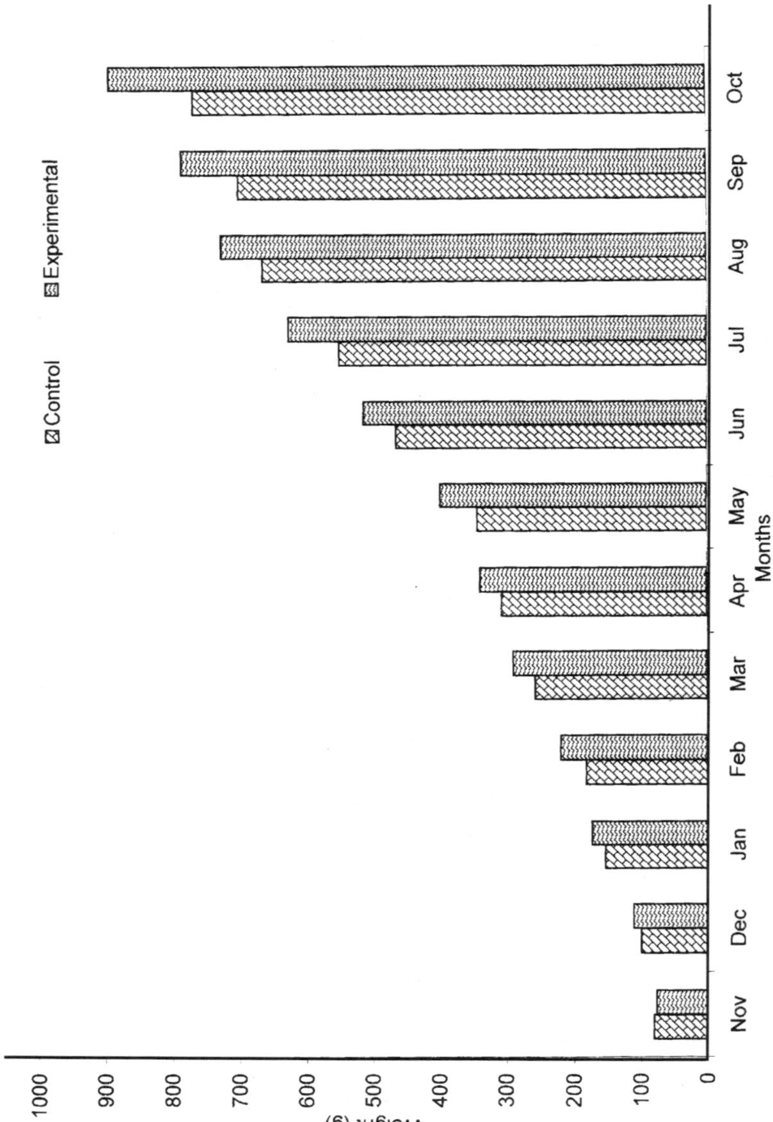

Figure 32: Growth of all Fishes Pooled

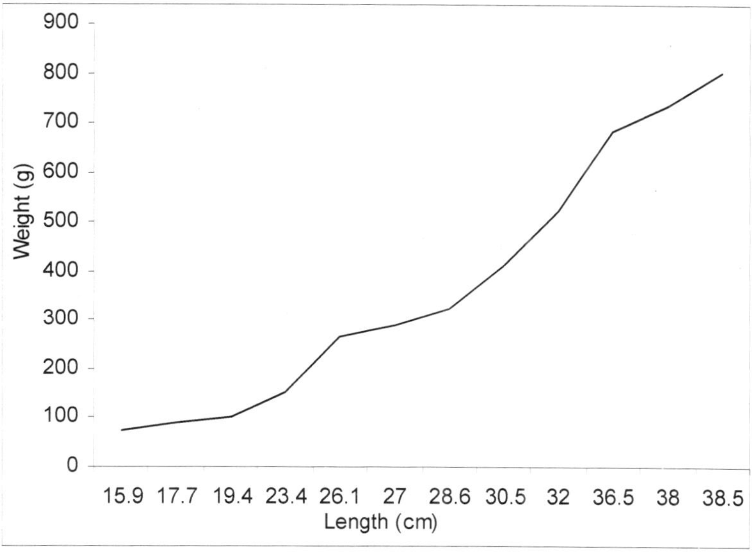

Figure 33: Length-weight Relationship of *Catla catla* in Control Pond

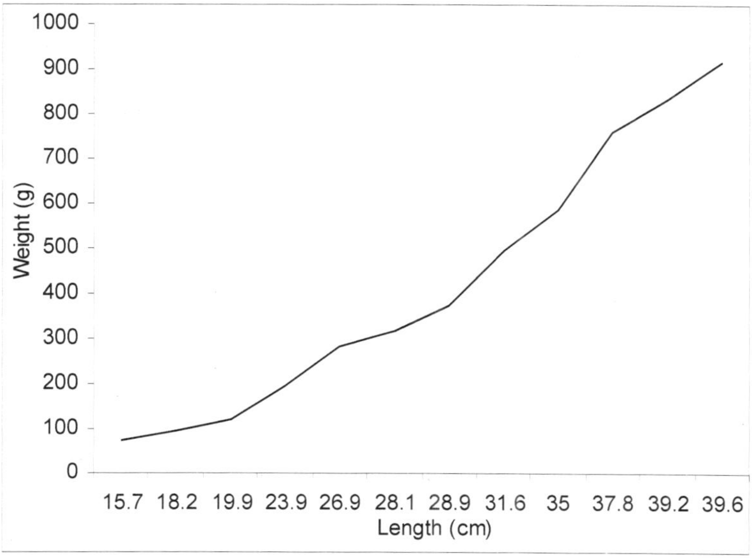

Figure 34: Length-weight Relationship of *Catla catla* in Experimental Pond

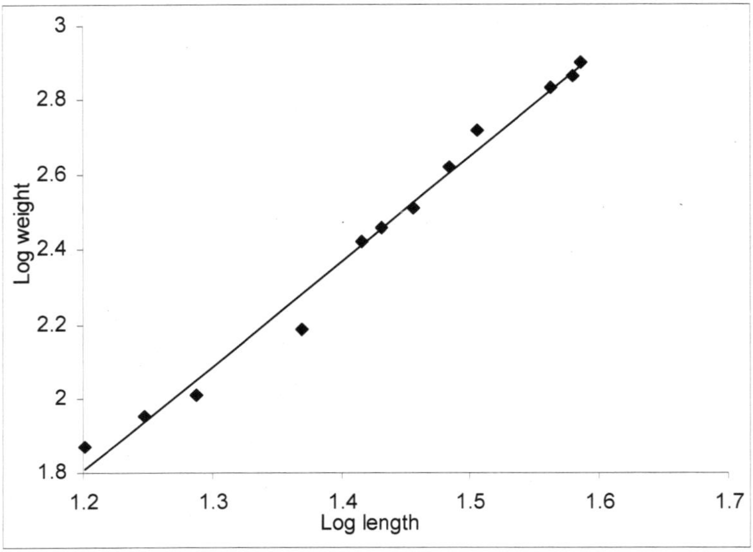

Figure 35: Logarithmic Length-weight Relationship of *Catla catla* in Control Pond

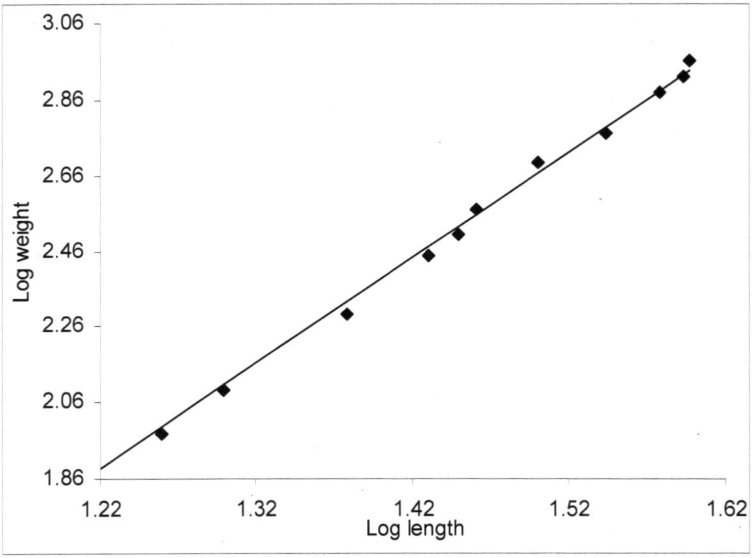

Figure 36: Logarithmic Length-weight Relationship of *Catla catla* in Experimental Pond

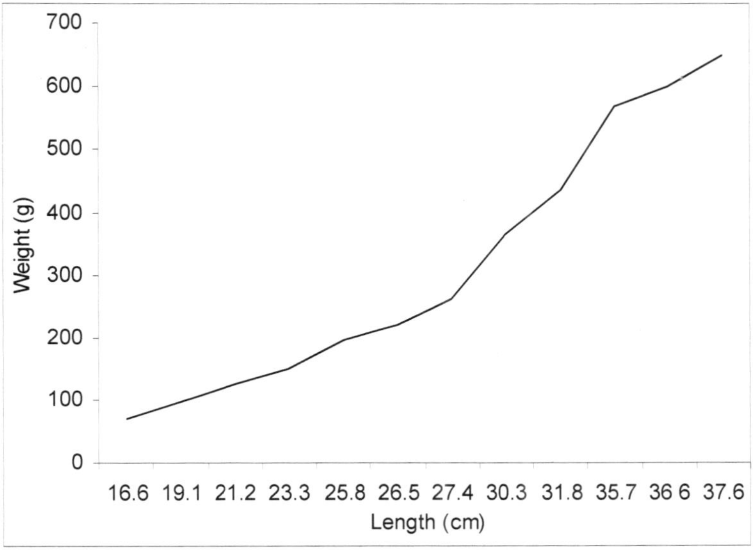

Figure 37: Length-weight Relationship of *Labeo rohita* in Control Pond

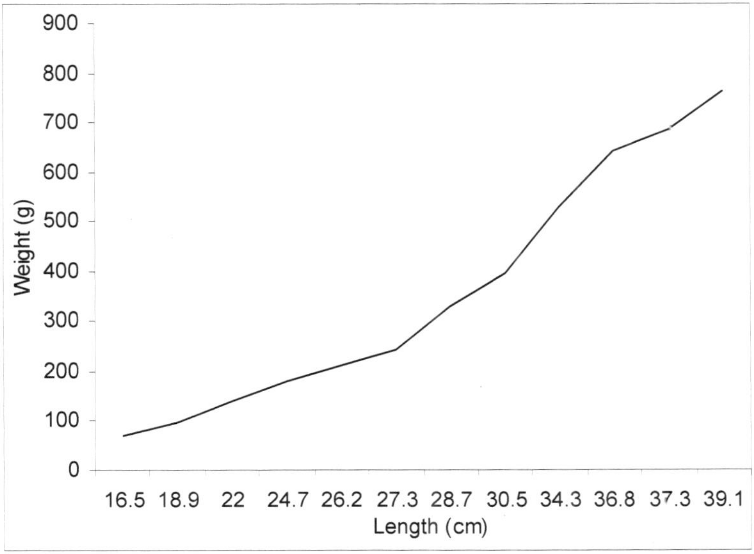

Figure 38: Length-weight Relationship of *Labeo rohita* in Experimental Pond

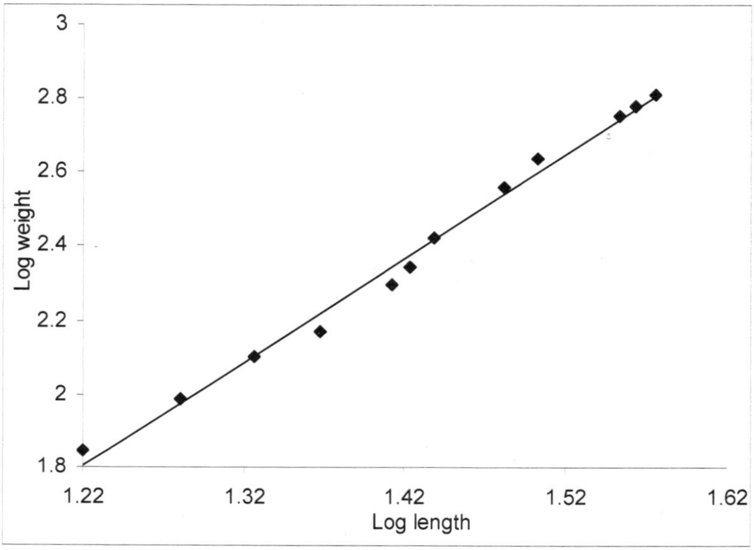

Figure 39: Logarithmic Length-weight Relationship of *Labeo rohita* in Control Pond

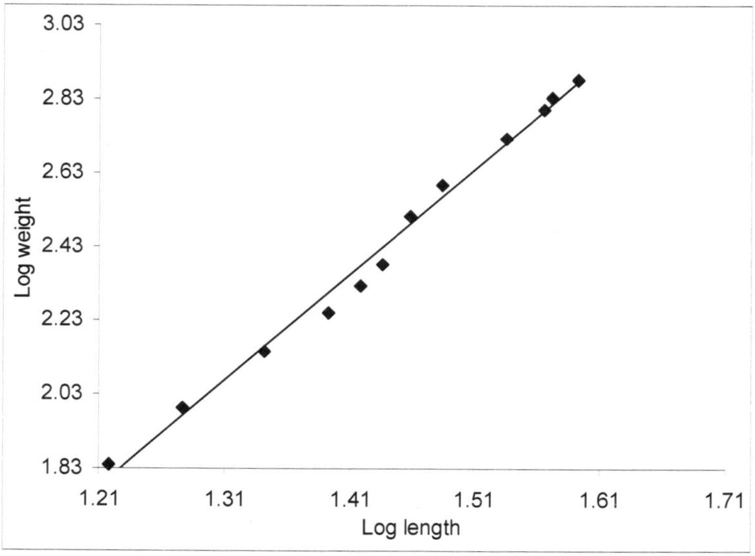

Figure 40: Logarithmic Length-weight Relationship of *Labeo rohita* in Experimental Pond

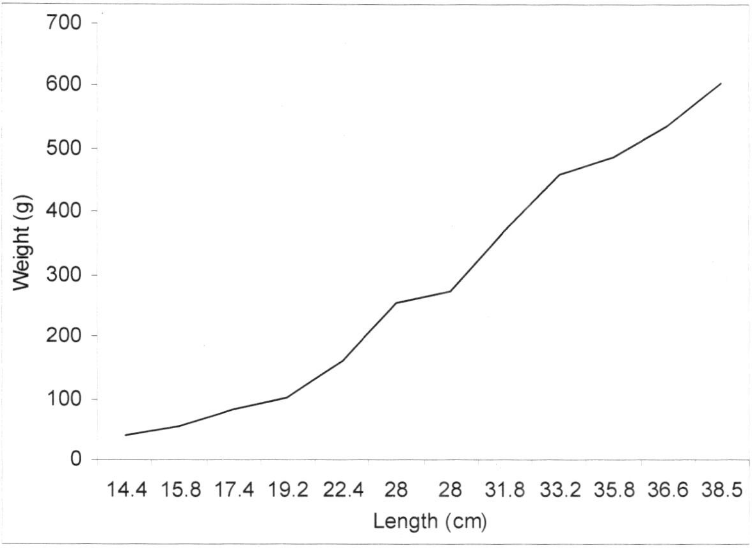

Figure 41: Length-weight Relationship of *Cirrhinus mrigala* in Control Pond

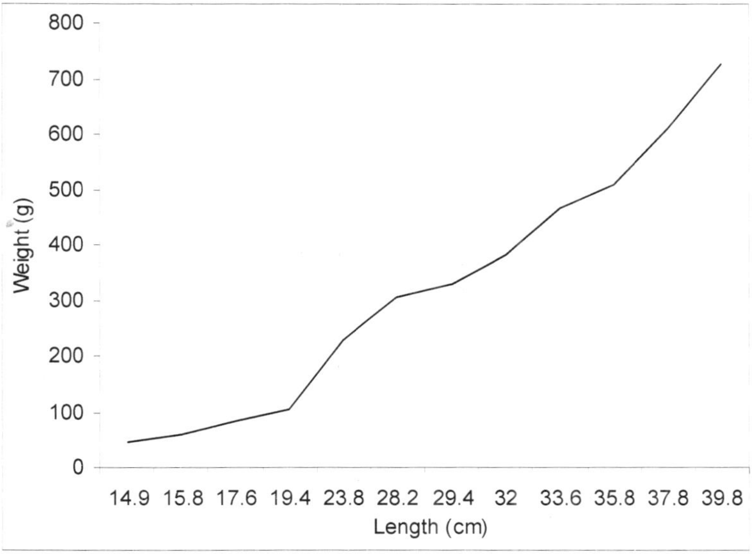

Figure 42: Length-weight Relationship of *Cirrhinus mrigala* in Experimental Pond

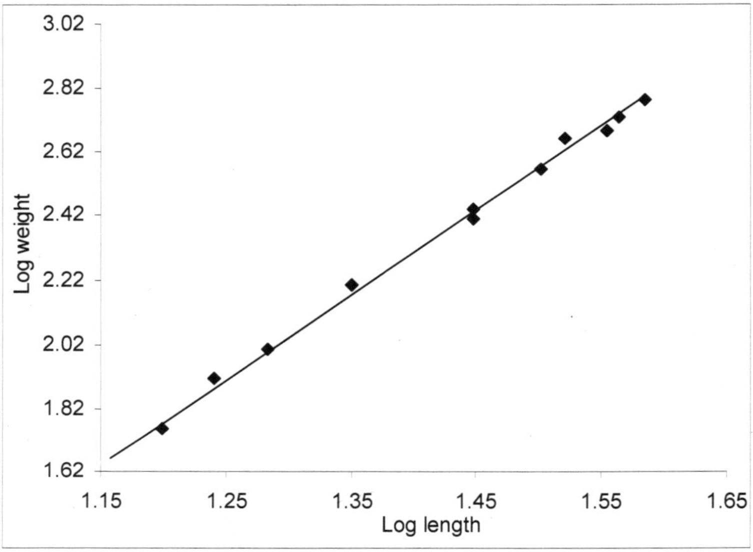

Figure 43: Logarithmic Length-weight Relationship of *Cirrhinus mrigala* in Control Pond

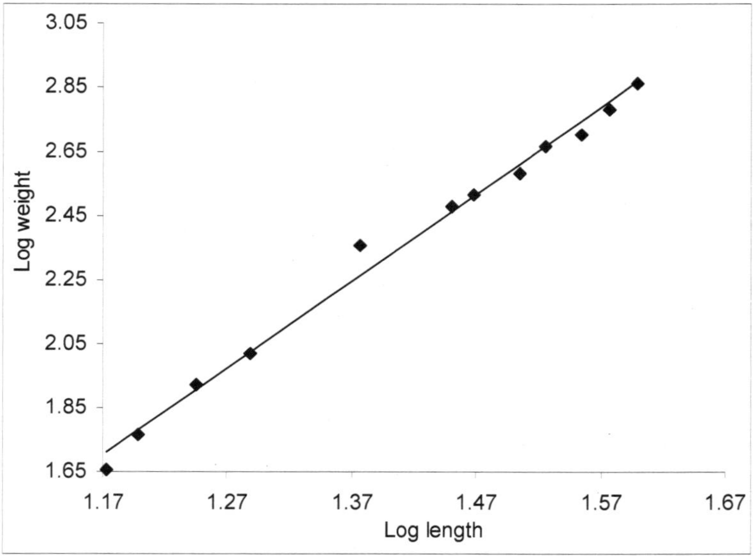

Figure 44: Logarithmic Length-weight Relationship of *Cirrhinus mrigala* in Experimental Pond

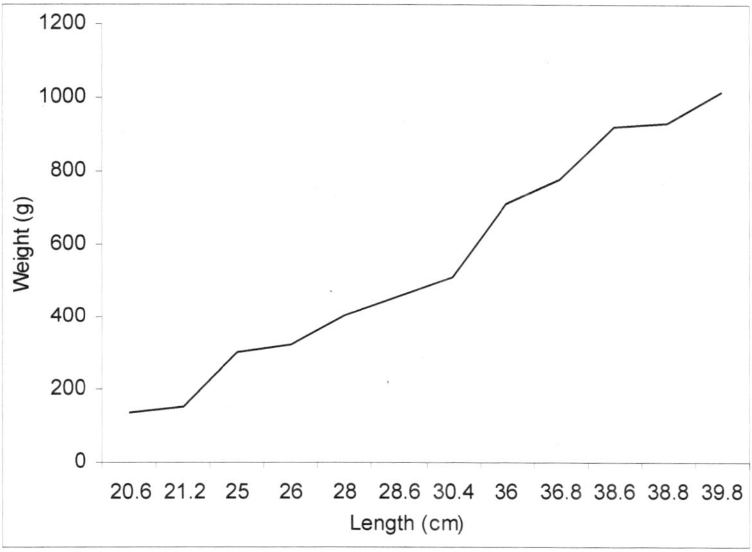

Figure 45: Length-weight Relationship of *Cyprinus carpio* in Control Pond

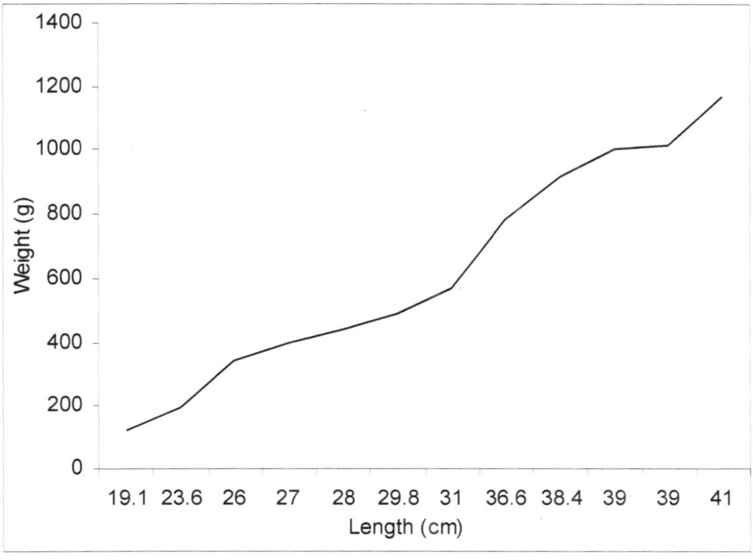

Figure 46: Length-weight Relationship of *Cyprinus carpio* in Experimental Pond

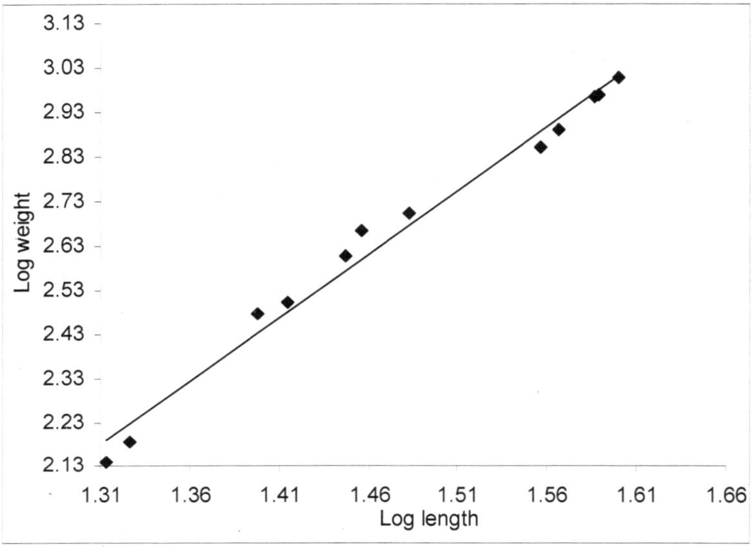

Figure 47: Logarithmic Length-weight Relationship of *Cyprinus carpio* in Control Pond

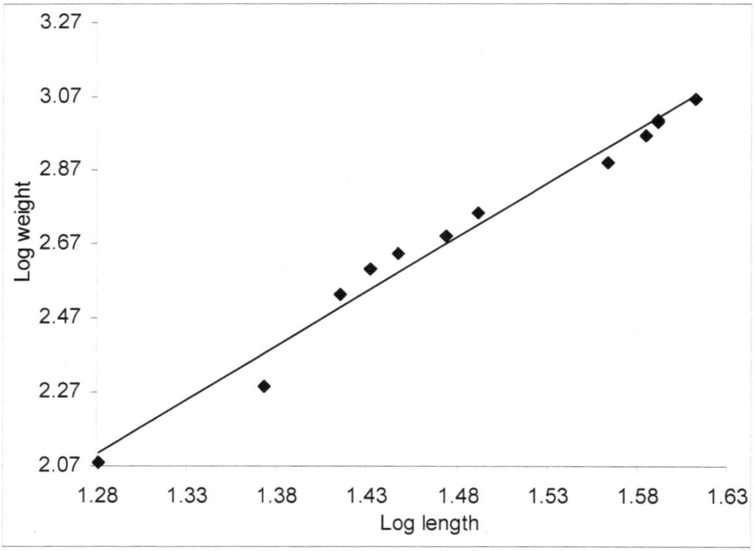

Figure 48: Logarithmic Length-weight Relationship of *Cyprinus carpio* in Experimental Pond

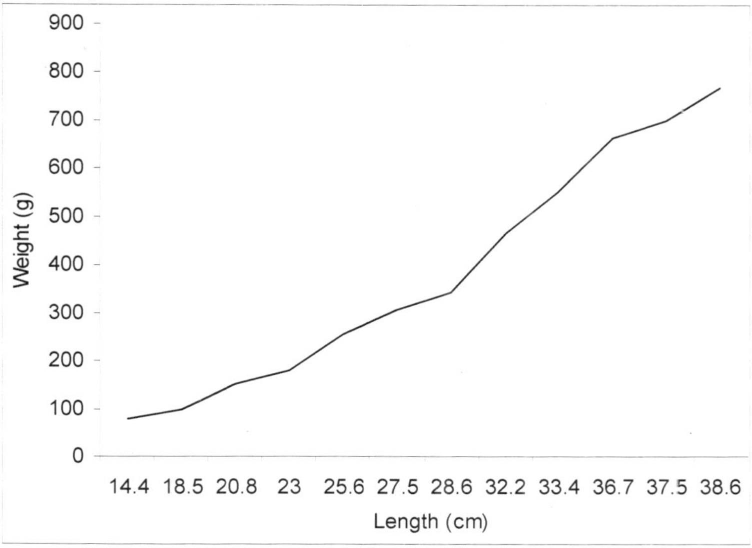

Figure 49: Length-weight Relationship of All Fishes Pooled in Control Pond

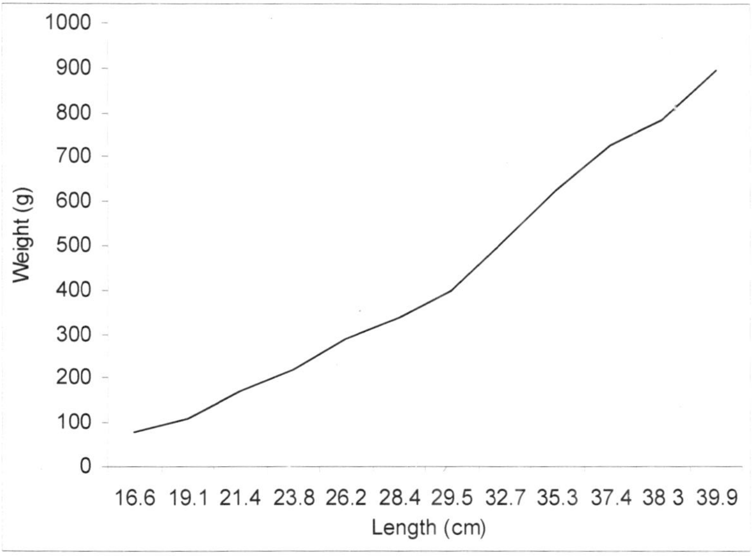

Figure 50: Length-weight Relationship of All Fishes Pooled in Experimental Pond

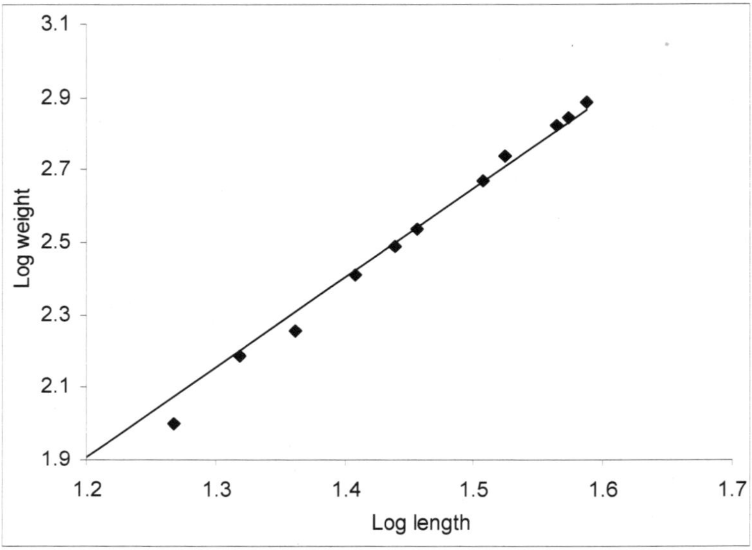

Figure 51: Logarithmic Length-weight Relationship of All Fishes Pooled in Control Pond

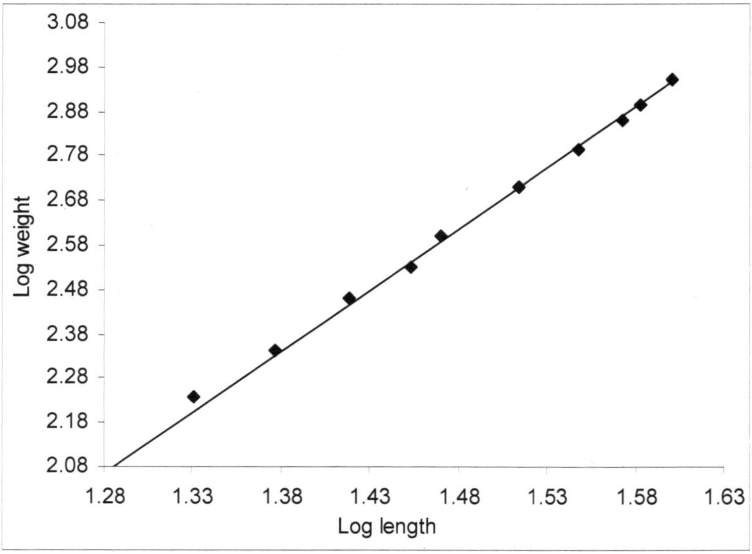

Figure 52: Logarithmic Length-weight Relationship of All Fishes Pooled in Experimental Pond

Labeo rohita, 0.9856 for *Cirrhinus mrigala*, 0.9925 for *Cyprinus carpio* and 0.97 for all fishes pooled in control pond (Tables 24, 25). The values of condition factor 'k' for all the fishes stocked in experimental and control pond are listed in Table 26. These value were in the range of 1.1515-1.6906 in experimental and 1.0613-1.6109 in control pond. The 'K' value for fishes pooled in experimental pond was 1.4055, while it was 1.3343 in control pond.

Table 23: Specific Growth Rate (SGR) in Different Fishes Stocked in Experimental and Control Pond

Species	*SGR*	
	Experimental Pond	*Control Pond*
Catla catla	0.6898	0.6505
Labeo rohita	0.6575	0.6105
Cirrhinus mrigala	0.7611	0.7356
Cyprinus carpio	0.6246	0.5483

Table 24: Length-Weight Relationship of Fish Reared in *Azolla*-pig-fish Experimental Pond

Species	*Exponential Equation*	*Logarithmic Equation*	*Correlation Coefficient 'r'*
Catla catla	$W = 0.0371\ L^{2.7794}$	Log W = 1.5059+2.7794 log L	0.9727
Labeo rohita	$W = 0.0236\ L^{2.8675}$	Log W = 1.6900+2.8675 log L	0.9728
Cirrhinus mrigala	$W = 0.0417\ L^{2.7121}$	Log W = 1.4733+2.7121 log L	0.9834
Cyprinus carpio	$W = 0.0266\ L^{2.9135}$	Log W = 1.6256+2.9135 log L	0.9877
All fishes pooled	$W = 0.0405\ L^{2.7523}$	Log W = 1.4571+2.7521 log L	0.9836

Nutritive value of fish at the end of experiment was assessed to analyze the samples of each species of each pond for moisture content, crude protein, crude fat and ash content on per cent basis. Values are presented in Table 27. Crude protein was slightly higher in all species of experimental pond than the fishes of control pond. Similarly, ash content and moisture was also higher in all species of experimental pond, but crude fat was higher in all fish species of control pond. Maximum protein was recorded in *Labeo rohita* fish

(18.14 per cent) but ash content was high in *Cyprinus carpio* (7.3 per cent). Maximum crude fat was 2.2 per cent in *Cyprinus carpio*. Mean values were computed as 75.97±2.6 per cent for moisture content, 17.99±0.12 per cent for crude protein, 1.48±0.33 per cent for crude fat and 6.75±0.42 per cent for ash content in experimental pond. In control pond mean values were computed as 74.71±1.28 per cent for moisture content, 17.68±0.32 per cent for crude protein, 1.95±0.24 per cent for crude fat and 5.95±0.39 for ash content.

Table 25: Length-Weight Relationship of Fish Reared in Pig-Fish Control Pond

Species	*Exponential Equation*	*Logarithmic Equation*	*Correlation Coefficient 'r'*
Catla catla	$W = 0.0313\,L^{2.8191}$	Log W = 1.5763+2.8191 log L	0.9743
Labeo rohita	$W = 0.0253\,L^{2.8213}$	Log W = 1.6368+2.8213 log L	0.9771
Cirrhinus mrigala	$W = 0.0469\,L^{2.6577}$	Log W = 1.4173+2.6577 log L	0.9856
Cyprinus carpio	$W = 0.0297\,L^{2.8672}$	Log W = 1.5741+2.8672 log L	0.9925
All fishes pooled	$W = 0.1018\,L^{2.4686}$	Log W = 1.0526+2.4686 log L	0.9753

Table 26: Condition Factor 'k' in Different Fishes Stocked in Experimental and Control Pond

Species	*Value of 'k'*	
	Experimental Pond	*Control Pond*
Catla catla	1.4786	1.4043
Labeo rohita	1.2747	1.2165
Cirrhinus mrigala	1.1515	1.0613
Cyprinus carpio	1.6906	1.6109
All fishes pooled	1.4055	1.3343

Pig Production

Observation on growth performance, excreta production, FCR and yield of pigs were taken for two lots of six months duration (November 2002-April 2003 and May 2003-October 2003), recorded data are presented in Table 28.

Table 27: Nutritive Value of Fish at the End of Experiment (in %)

Species	*Experimental*				*Control*			
	MO	*CP*	*CF*	*AS*	*MO*	*CP*	*CF*	*AS*
Catla catla	79.40	17.98	1.8	6.3	73.97	17.12	2.1	5.6
Labeo rohita	75.89	18.14	1.1	6.8	76.05	17.85	1.7	5.9
Cirrhinus mrigala	75.39	17.86	1.3	6.6	75.49	17.69	1.8	5.8
Cyprinus carpio	73.19	17.98	1.7	7.3	73.34	17.67	2.2	6.5
Mean values	75.97±2.6	17.99±0.12	1.48±0.33	6.75±0.42	74.71±1.28	17.58±0.32	1.95±0.24	5.95±0.39

MO: Moisture; CP: Crude protein; CF: Crude fat; As: Ash content.

Plate 14: Piglets with Experimental Pond in First Lot

Plate 15: Piglets with Experimental Pond in Second Lot

Plate 16: Six Months Reared Adult Pig

Average gain in live weight of pigs was slightly higher in batch of pigs fed with azolla supplemented diet than batch of pigs fed with commercial diet in both lots. Average final weight of first batch (fed with azolla supplemented diet) during first lot was recorded as 77.5 kg with average gain in weight as 64.45 kg. It was recorded as 76.00 kg and 63.05 kg for second lot of same batch. Average final weight of second batch (fed with commercial diet) was 74.75 kg with average gain in weight as 62.30 kg for first lot and 75.2 kg and 61-95 kg for second lot. Total live weight of pigs reared in both lots with azolla supplemented diet was 845 kg and it was recorded as 824.5 kg for pigs reared with commercial feed. Average feed conversion ratio was computed as 5.62 and 5.7, 5.76 and 5.92 for first and second group of pigs in first and second lot, respectively.

A separate experiment was designed to estimate excreta/day/ pig. Excreta (dung + urine + spilled over feed) of 24 hrs. was collected from animal of different age group and average production of excreta/ animal/ day was calculated (Table 29). On average, 794.73 g waste was produced by an animal in a day and 143.05 kg in 6 months.

Hence, 1573.75 kg. piggery effluent was added in each pond. Spilled over feed was negligible for both groups.

Observation recorded on various parameters (nitrogen, Phosphorus, Potassium and Moisture content) for wastes of pigs, fed with common diet (RPW), wastes of pigs, fed with azolla supplemented diet (SPW) and azolla treated pig slurry (APW). Data are presented in Table 30. Maximum per cent of nitrogen was observed in APW (2.02 per cent) followed by SPW (1.88 per cent) and RPW (1.78 per cent). Phosphorus was maximum in SPW (0.60 per cent) followed by RPW (0.45 per cent) and APW (0.40 per cent). Minimum moisture was recorded in RPW (69.5 per cent), while it was maximum (72 per cent) in APW.

Table 28: Growth Performance of Pigs Fed with Commercial Diet and *Azolla* Supplemented Diet

Parameters	*First Lot (6 Months)*		*Second Lot (6 Months)*	
	Control (PA)	*Experimental (PB)*	*Control (PA)*	*Experimental (PB)*
Nos. of pigs	6	6	5	5
Av. initial wt. (kg)	12.45	13.05	13.25	12.95
Av. final wt. (kg)	74.75	77.5	75.2	76.00
Av. gain in wt. (kg)	62.30	64.45	61.95	63.05
Av. feed consumption (kg)	359	362	367	365
Av. feed conversion ratio	5.76	5.62	5.92	5.79

PA: Pigs fed with commercial dict.
PB: Pigs fed with 20 per cent dried azolla + 80 per cent commercial feed.

Table 29: Estimation of Wastes Released by Pig (Per Day)

Age of Pig (Month)	*Av. Weight of Pig (kg.)*	*Feed Gain (gm)*	*Wastes* Released (gm)*
3	12.1	400	280.6
4	21.4	850	685.2
8	64.0	2000	1418.4

Average waste per day (gm) 794.73

* Including dung + urine + spilled feed.

Wastes released by a pig in 6 months = 143.05 kg.

Commercial pig feed (pig mash) and azolla supplemented pig feed was analysed for crude protein, crude fat and crude fibre and data are presented in Table 31. Values of all nutrient are higher in azolla supplemented feed. They were obtained as 17.2 and 16.0 for crude protein, 6.1 and 6.0 for crude fat and 14.2 and 12.8 for crude fibre in azolla supplemented feed and common feed, respectively.

Table 30: Proximate Composition of Nutrients in Pigdung (%)

Parameters	*RPW (Azolla Diet)*	*RPW (Common Diet)*	*APW*
Nitrogen	1.88	1.78	2.02
Phosphorus	0.60	0.45	0.40
Potassium	0.35	0.40	0.35
Moisture	70.00	69.50	72.00

Table 31: Proximate Composition of Pig Feed

Nutrients	*Commercial Feed*	*Commercial Feed (80%) + Azolla (20%)*
Crude protein	16.0	17.2
Crude fat	6.0	6.1
Crude fibre	12.8	14.2

Azolla Production

Azolla was cultivated in waste digestion pond and harvested biomass was used in pig diet. Observation for fresh biomass production of azolla and doubling time was recorded (Table 32). Average net production was 41.7 kg/week in 24m^2 and 1.74 kg/m^2/week with 3 days doubling time. Average moisture content of fresh azolla biomass was recorded as 87.6 per cent after sun drying and 94.2 per cent after oven drying (80°C for 8 hours). Growth performance of four varieties were tested in four earthen raceways (400×100×40 cm) fertilized with piggery effluent and found that *A. pinnata* asitica and *A. microphylla* were suitable for this integration *A. filiculoides* and *A. rubra* showed negligible survival and growth. Brown raddish colouration was observed during winter. The production was adversely affected by hot environment (>38°C) during May-June and by cold (<20°C) during December and January. Shade over the pond and down level of pond water (20 cm from ground level) resulted as normal growth of azolla during adverse conditions. Samples of oven dried azolla were analysed for proximate composition and data were recorded in Table 33. According to the perusal of data, azolla contain 94.2 per cent moisture, 22.0 per cent crude protein, 6.1 per cent crude fat and 18.5 per cent crude fibre on dry weight basis 9 kg/ week fresh azolla biomas is required to supplement the diet of a pig at 20 per cent level, which can be produced in 5m^2 area.

Table 32: Yield of *Azolla* in Pig Effluent Treated Pond (Pond Size 4 × 6 m)

Parameters	*Quantity (kg)*			
	0 Week	*1st Week*	*2nd Week*	*3rd Week*
Initial wet wt. of azolla inoculated	10	10	10	10
Wt. of harvested biomass	-	52.2	49.8	53.1
Net production	-	42.2	39.8	43.1
Moisture content (per cent)	-	93.2	94.4	94.6

Table 33: Proximate Composition of *Azolla*

Parameters	*(%) of Dry Weight*
Moisture	94.2
Crude protein	22.0
Crude fat	6.1
Crude fibre	18.5

Table 34: Nutrient Contents in Piggery Effluent and their Recycling in Fish Pond (0.12 ha.) during Experimental Period

Nutrient	*Quantity Added (kg.)*		*Calculated Quantity/ha./ Year (kg.)*	
	Experimental	*Control*	*Experimental*	*Control*
Nitrogen	29.583	28.009	246.52	233.41
Phosphorus	9.44	7.08	78.68	59.00

SPW: Waste of pigs fed with *azolla* supplemented diet; RPW: Waste of pigs fed with commercial diet; APW: *Azolla* treated pig waste.

Plate 17: Production of *Azolla pinnata*

Plate 18: Plant Step of *Azolla pinnata*

Plate 19: Plant Step of *Azolla microphylla*

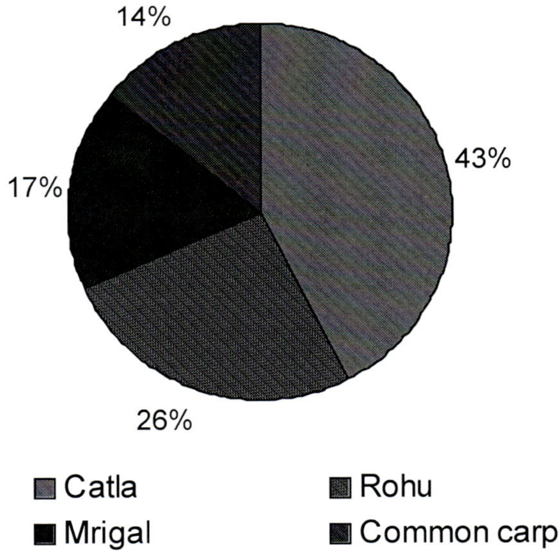

Figure 53: Species-wise Contribution in Total Fish Production of Control Pond

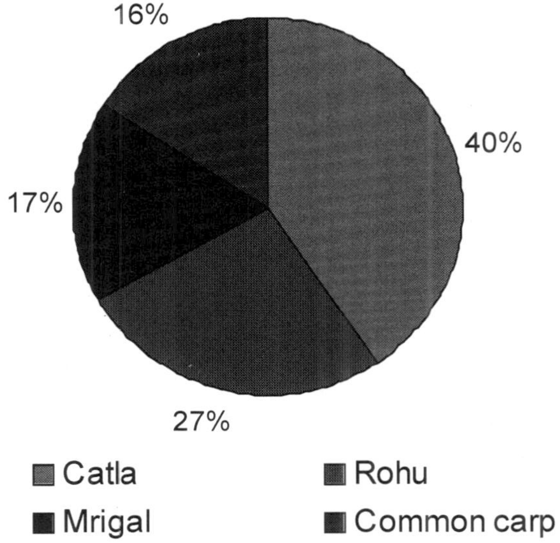

Figure 54: Species-wise Contribution in Total Fish Production of Control Pond

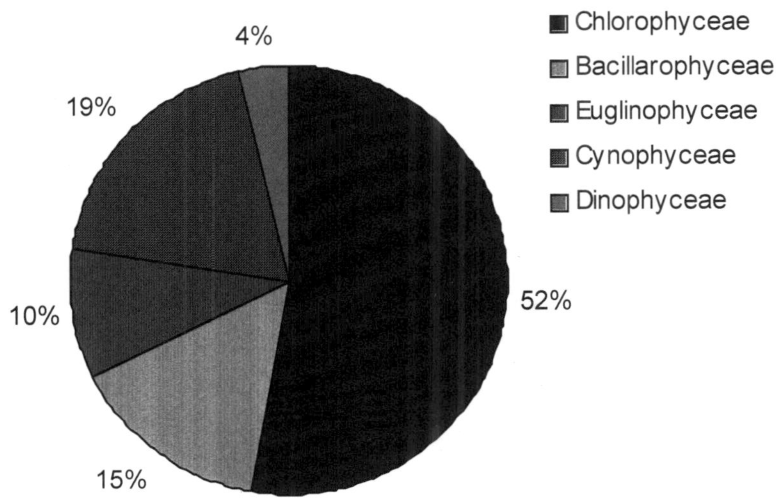

Figure 55: Composition of Phytoplankton in Control Pond

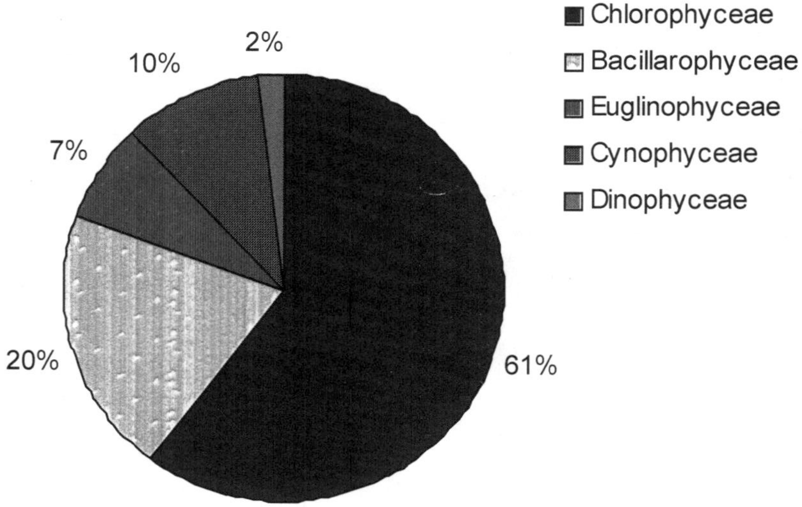

Figure 56: Composition of Phytoplankton in Experimental Pond

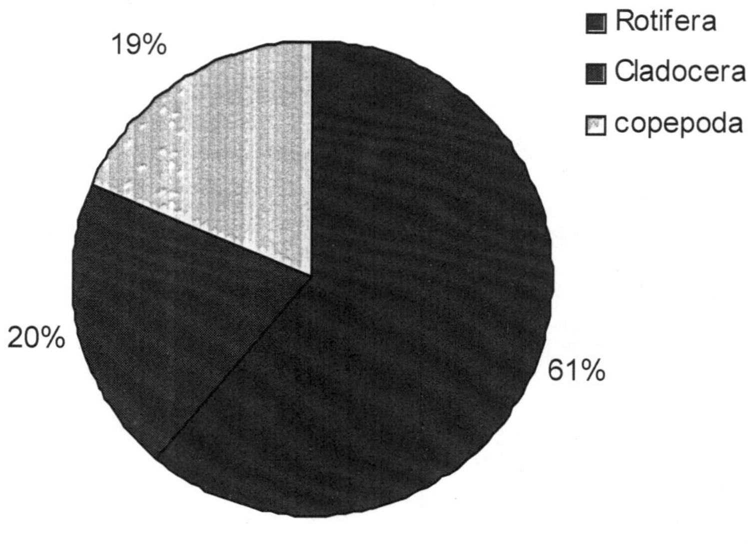

Figure 57: Composition of Zooplankton in Control Pond

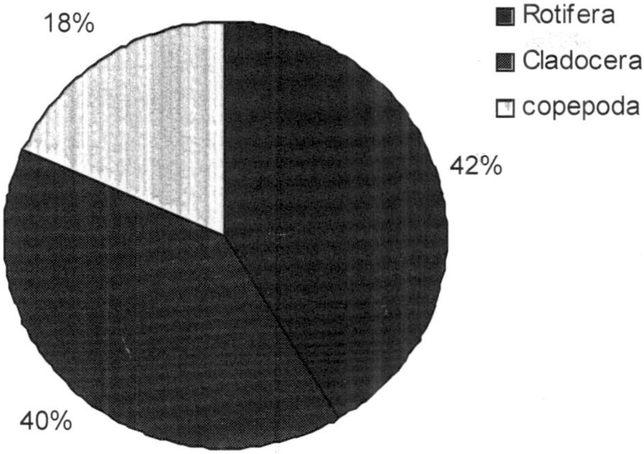

Figure 58: Composition of Zooplankton in Experimental Pond

Plate 20: *Azolla* Production in Earthen Raceway

Recycling of Waste and Nitrogen Budget

Details of recycling of pig excreta are presented in Table 22. Details of the manurial value of pig excreta are presented in Table 10. A total of 29.583kg nitrogen and 9.44 kg phosphorus was added in experimental pond in the form of pig effluent at the rate of 246.52 kg N_2 /ha/yr. and 78.68 kg Phosphorus/ha/yr. it was 28.009 kg nitrogen and 7.08 kg phosphorus at the rate of 233.41 kg nitrogen/ ha/yr and 59.00 kg phosphorus/ha/yr. in control pond. A complete nitrogen budget of each pond is presented in Table 35.

Nitrogen fixation by azolla was quantified to calculate the total budget of nitrogen assuming the same untraced quantity in both ponds. It was not determined by any standard method. Simple calculated quantity of fixed nitrogen by azolla was 5.093 kg with a fixing rate 0.88 mg/day/g fresh azolla biomass. At the end of experiment the retention of added nitrogen in different components were as follows: pond water 3.6 per cent, pond sediment 17.8 per cent plankton 1.9 per cent, benthos 1.2 per cent, fish 52.9 per cent azolla 12.3 per cent and 10.3 per cent as untraced (Table 35).

Table 35: Recycling of Nitrogen in Experimental (E) and Control (C) Fish Pond

Parameters	*Pond*	
	E	*C*
Nitrogen added in pond		
(a) As pig dung (g)	29583	28009
(b) Fixation by azolla	5093	-
Total nitrogen added (g)	34676	28009
Nitrogen retained		
(a) Pond water (g)	1237	514
(b) Pond soil (g)	6183	8246
(c) Plankton (g)	649	973
(d) Benthos (g)	439	165
(e) Fish (g)	18350	1455
(f) Azolla (g)	4257	-
(g) Untraced (g)	3561	3561

Table 36(a): Total Input Cost of Experimental and Control Units
Experimental: Integrated Azolla-Pig-Fish unit
Control: Integrated Pig-Fish Unit

Components	*Items*	*Unit*	*Experimental*			*Control*		
			Qt. / No.	*Rate (Rs.)*	*Cost (Rs.)*	*Qt. / No.*	*Rate (Rs.)*	*Cost (Rs.)*
A. Fish	Pond preparation	–	–	LS*	1000	–	LS	1000
	Lime	Kg.	36	2	72	36	2	72
	Carry over Seed	No.	960	2/seed	1920	960	2/ seed	1920
	Harvesting	–	–	LS	200	–	LS	200
	Water rent	–	–	LS	200	–	LS	200
B. Livestock	Pig house	–	–	LS	500	–	LS	500
	Piglets	No.	11	480	5280	11	480	5280
	Pig feed	Qtl.	32	415	13280	40	415	16600
	Medicine	–	–	LS	275	–	LS	275
	Miscellaneous	–	–	LS	235	–	LS	235
C. Azolla	Pond preparation	–	–	LS	200	–	–	–
	Pesticides	–	–	LS	75	–	–	–
D. Labour	Wage	Month (Rs.)	12	250	3000	12	250	3000
	Total input cost				**26237**			**29282**

*: Lumpsum.

Table 36(b): Comparative Production Economics of Experimental and Control Unit
Pond area/unit = 0.12 ha, Pig unit = 2 nos., Azolla unit = 24 m^2

		Production and Return			
Community	*Price Rate (Rs.)*	*Experimental*		*Control*	
		Production (kg.)	*Sale price (Rs.)*	*Production (kg.)*	*Sale price (Rs.)*
Pig (live weight)	29.5	845	24927.50	824.5	24322.75
Fish	34	637.8	21685.00	531.2	18060.80
Total return (Rs.)			46612.50		42383.55
Net return (Rs.)			20375.50		13101.55
Percent return to variable cost			77.66		44.74

Plate 21: Plankton Estimation

Economics of Integrated Fish Production

Total input cost of experimental and control unit was recorded in Table 36. The total input cost of experimental unit, integrated with piggery as well as with azolla production cum waste digestion unit was Rs. 26237/-. It was Rs. 29282/- for non-azolla integrated unit. Details of production and return is presented in Table 37. In

Plate 22: Appearance of Algal Bloom

experimental unit, the percent return over variable cost was 77.66 and it was 44.74 for control unit, which was significant at 0.05 level.

Chapter 6

Discussion

Integrated livestock fish farming system is a proven environmentally sustainable and economically viable technology that encompasses rational utilization of available resources. Different forms of integrated livestock fish farming *viz.* pig-fish, poultry-fish, duck-fish etc. have been evolved and popularized in India (Sharma *et al.*, 1985). Efforts are being made to improvise the technology by way of multiplication of production potentiality and minimization of risk factor through incorporation of more components into the system (Sharma, 1989). The present study was undertaking on the feasibility of an integrated livestock fish farming system in which three components *viz.* pig, fish and azolla were incorporated and azolla was utilized to partially supplement the nutritional requirement of pigs. For comparison, control set of pig-fish integration was run following standard specification for semi intensive composite carp culture (Anon, 1993).

The bottom sediments and water of fish ponds play important role in the productivity. Various physical and chemical factors of pond water and soil determine its biological productivity. The nutrients are stored in the bottom soil. Mineralization and exchange of the nutrients also occur in the pond mud. The amount of organic matter present in the bottom is index of its productivity. The physical and chemical characteristics of water are reflection of the bottom sediments. The physical factors of water such as temperature, light penetration, depth, water current, turbidity etc play major role in the overall metabolism of water bodies. Chemical factors like dissolved gases, buffer system and nutrients are also of great importance to all living organisms. The understanding of the physical and chemical properties of soil and water enables best possible mode of exploitation of pond water from fish culture point of view.

Chemical characteristics of pond bottom mud greatly influence the supply of nutrients for primary production through mineralization of decomposing organic debris creating suitable condition for existence of aquatic life. The soil of the fish ponds are continually altered due to decomposition of organic matter added in the form of manure. The nature of pond sediment of the water bodies and its chemical composition influence the aquatic primary productivity in the same way as they affect the terrestrial vegetation.

In spite of about facts, very little attention has been paid towards the study of pond soil (Banerjea and Ghosh, 1970; Saha, 1985). The pH of pond soil plays crucial rate in the availability of essential nutrients. Banerjea (1967) and Jhingran (1988 and 1991) have reported that moderate alkaline nature of the bottom soil (pH-7.2-7.4) is indicative of optimal conditions for fish production. In present experiment soil pH were slightly alkaline (7.4-7.6), suggesting the productive nature of pond bottom. The value of organic carbon registered increasing trend towards the end of experiment with a peak in February. As the process of decomposition is temperature depended, it slows down during the winter months leading to the accumulation of organic carbon in pond soil. In present experiment, slightly elevated concentration of organic carbon (54±1.385 per cent) was found during summer months. Similar results were reported by Saha (1985). Pig excreta was the main source of organic carbon in the ponds, besides the autochthonous organic matter. Chatterjee and Saha (2000) also recorded significant increase ($P<0.01$) in the organic carbon content of pond soil having 0.45 per cent and 1.33 per cent organic carbon content with the application of organic manures at different doses. Nitrogen level which was minimum during the initial period of experiment gradually increased attaining highest values at the end of experiment. The increase in soil nitrogen content towards end of the experiment may be attributed to the constant deposition of excess nitrogenous substances to the pond bottom. Seasonal variations in the values of nitrogen in pond soil due to the addition of nitrogenous compounds was also reported by Saha (1985) in two natural fresh water pond.

Banerjia (1967) has attempted to correlate fish production with the available nitrogen in soil of fresh water fish ponds and recommended nitrogen in the range of 25-75 mg/100 g soil as relatively more favorable for fish production. Considerably good nitrogen values (30.6-66.0 mg/100g and 29.47-59.37 mg/100g) obtained in the soil of both ponds. The present study points out that the total amount of available nitrogen in soil was not affected by cultivating azolla in integrated pig-fish culture system as mean values (49.1±10.48 and 48.97±10.29 mg/100g soil) are almost similar in both ponds and these data are non-significantly different at 0.05 level.

Phosphorus is one of the most important nutrient limiting primary productivity in most of the fresh water system. Soil acts as a reservoir of phosphorus which is released slowly to the overlying water and made available to the primary production (Boyd, 1982). According to Banerjea (1967), Sreenivasan (1967) and Jhingran (1988, 1989), the available phosphorus content of <3mg/100 g, 3-6 mg/100 g and >6mg/100 g soil are indicative of poor, moderate and high productivity, respectively. In the present study, the phosphorus content of experimental pond was in the range of 2.21-5.99 mg/100g, with mean value of 4.94±1.10 mg/100g soil, which was lower than the control (mean value 5.87±0.85 mg/100g soil) pond may be due to the uptake by azolla biomass. Hence, a significant difference was observed in the available phosphorus content of both ponds. Watanabe *et al.* (1980) reported the one kg uptake of phosphorus by azolla for fixation of 1.9-2.5 kg of nitrogen. Manimaran *et al.* (1997) also reported depletion of phosphorus from aqueous media cultivation of azolla. However, the existing values of phosphorus content reflect the good productive status of soil. The value of phosphorus concentration recorded in the present study are comparable to those obtained by Nath *et al.* (1994) in semi-intensive fish culture ponds. All chemical parameters of pond soil expect pH showed significant positive correlation with GPP, while pH showed negative correlation.

Study of the physico-chemical parameters of an aquatic ecosystem is basic for understanding its biological productivity.

Although each factor plays its individual role but it is the synergistic effect of various parameters which determines the composition and productivity of the flora and fauna. Conducive range of these factors is essential for obtaining optimum fish production.

Water temperature significantly affects the biological productivity of ponds by controlling chemical and metabolic reactions (Hutchinson, 1957). Low temperature reduces the metabolic activity of the organism (Khanna, 1993). Changes in temperature of the water body can influence the speed and course of decomposition of organic matter (Karause, 1962). In the present study, water temperature was found in the range of 14.7-32.0°C. Boyd and Pillai (1984) reported that the water temperature upto 32°C in tropical fish ponds does not seem to adversely affect on fish production. In present study a positive correlation was found between water temperature and GPP.

The hydrogen ion concentration (pH) of pond water is considered as an index of environmental conditions. It affects the chemical and biochemical reactions and controls the activities and distribution of aquatic fauna and flora (Verma and Shukla, 1969). According to Boyd and Pillai (1984) better fish production could be possible in pond water with pH value ranging between 6.5-9.0. In present study, the pH values varied from 6.9-8.0 in experimental pond and 6.8-7 in control pond, which indicates the favorable condition of productivity. Gradual decrease in pH value just after the fertilization supports the report of Alabaster and Lloyd (1980); Sharma and Saini (1992).

Dissolved oxygen is the most crucial factor for the growth and survival of fish. The optimum concentration of DO in pond waters is 6-9 ppm (Boyd, 1982). A minimum acceptable level is considered to be 5ppm (Eills, 1993). In pond water DO mainly evolves through the process of photosynthesis by phytoplankton. Inspite of rich planktonic organism (2.16±1.07 and 1.97±0.95 mg/50 litre) in pond water, dissolved oxygen remained in sub optimal level (5.41±0.77 mg/l and 4.83±0.35 mg/l). The low level of dissolved oxygen in these ponds, especially in winter months, reflects the richness of organic matter, which consumes large amount of dissolved oxygen in the process of decomposition. A significant difference ($P<0.05$) was observed between dissolved oxygen level of experimental (5.41±0.77 mg/l) and control (4.83±0.35 mg/l) which may be due to the partial decomposition of organic matter in stabilization pond prior to release into experimental fish pond. Das (1987) also observed decrease in DO when nutrient level increases due to application of manure. A significant negative correlation was found between DO and free CO_2. Similar correlation was observed by Singh and Sharma (1999). A significant positive correlation was found between DO and GPP in both ponds.

Presence of CO_2 is essential for photosynthesis but its excess is harmful to aquatic animals. Free CO_2 appeared in moderate concentration in the experimental pond during almost entire period of study. The values were relatively lower (0.9 mg/l) during summer and higher (9.2 mg/l) in winter. Baruha *et al.* (1998) also registered the same pattern. Comparatively greater utilization of CO_2 in photosynthesis may be the reason for lower values of free CO_2 in summer, whereas the increased value of free CO_2 during winter could be attributed to the higher rate of respiration of aquatic fauna than photosynthetic activity of aquatic flora (Shastri and Malik, 1992). Present study also reflected negative correlation of free CO_2 with pH of pond water and GPP.

Total alkalinity fluctuated in experimental pond with generally higher (91-177 ppm) than the range (100-120 ppm) suggested by Tripathi, 1982 for optimal fish production. The average lower value in control pond (146.67±22.6) than in experimental pond (155.3±24.7 ppm) could be attributed to higher release of free CO_2 in decomposition of organic matter (pig dung). The accumulation of free CO_2 reduces the pH of water as it was higher in control pond. In both ponds a significant positive correlation was found between alkalinity and GPP.

Salinity also affects the primary production in fresh water pond. In present study, salt concentration was always below (0.4-2.2 ppt) throughout study period as the critical limit suggested by Ghose *et al.*, 1973. The most important source of salt content was under ground brackish water of this area, which may entered into the pond with rain runoff water of catchment area of the ponds. This area was irrigated by this under ground brackish water for mustard crop. It's concentration was found higher during summer months (2.2 ppt) due to high evaporation of water from the pond and addition from piggery unit, where some brackish water was used for washing purpose during scarcity of fresh water in summer months. In present study, significantly positive correlation of salinity was found with GPP.

Continuous fluctuation was observed in the values of total dissolved solids (180-298 mg/l), which was positively correlated with GPP. Mean values of TDS in both ponds (241.92±39.98 and 248.92±46.43 mg/l) are non-significantly different ($P<0.05$). The range of TDS in present investigation supports the recommended level suggested by Boyd (1974).

In fish culture systems ammonia appears in water mainly from ammonification of organic matter by hetrotrophic bacteria and deamination and transmination of catabolic products of organic nitrogen ingested and assimilated by the cultured animals (Gopal Krishnan *et al.*, 1997). The mean values (0.14±0.08 and 0.11±0.06 mg/l) of ammonia in both ponds were always lower than the maximum acceptable level (1.0 ppm) as recommended by Boyd and Pillai (1984). Nitrite and nitrate concentration play an important role in primary production. The most important source of these elements are biological oxidation of organic manure. The atmospheric nitrogen fixed as ammonia by azolla was also be a significant source of nitrite and nitrate in experimental pond water. The fixed nitrogen in the form of ammonia is oxidized in to nitrite and nitrate by bacteria present in the cavity of lower lobs of azolla leaves. The concentration of nitrite (0.05±0.02mg/l) higher in control pond. But nitrate was high (0.78±0.32 mg/l) in experimental pond. In winter season the concentration of ammonia nitrogen (0.01 mg/l), nitrite nitrogen (0.018 mg/l) and nitrate nitrogen (0.07 mg/l) was minimum, probably due to reduced rate of decomposition of organic matter. In summer season, concentration of nitrite nitrogen (0.068 mg/l), and nitrate nitrogen (1.22 mg/l) was maximum, but ammonia nitrogen (0.241) was maximum during spring season in both pond. Singh (2002) also reported low concentration of ammonia nitrogen during the month of December in duck-fish integrated pond and high concentration during the month of June. Contrary to the above, increased values of ammonia as high as 2.0-2.4 ppm were estimated by Akpan and Okafor (1997) in a pond treated with poultry manure, during the month of December and January. In total nitrogen, nitrate nitrogen occurred in very high concentration, while ammonia nitrogen ranked next. Similar results were reported by Padmavati *et al.* (1997). Values of ammonia nitrogen and nitrate nitrogen show significant correlation with GPP.

Phosphorus is one of the major limiting nutrient for phytoplankton (Boyd, 1982). Since, it was present in very low concentrations in pond water, it was supplemented through the application of manure concentration of phosphorus in ponds exhibited bimodal pattern of fluctuation during the study period. The first peak was recorded in the month of January and second in April. Availability of phosphorus decreases from April onwards, owing to assimilation by increasing phytoplankton population. Comparatively lower (0.23±0.08 mg/l) concentration of phosphorus in experimental pond than in control pond (0.30±0.13 mg/l) indicates its retention by azolla (Watanable *et al.*, 1980). But it showed optimum level (0.05-2.0 ppm) as recommended by Benerjea (1967). A significant positive correlation was observed between the phosphorus content and GPP. The biological oxygen demand (BOD) is a measure of the amount of oxygen required by micro organism to decompose the organic matter in a water body (Boyd, 1978). The property has a direct bearing on the oxygen balance of water; particularly when the environment is rich in organic materials. Complete decomposition of organic matter require a long time and hence 5 day incubation period has been suggested for most practical purposes (APHA, 1971). The BOD value thus obtained is usually termed as BOD_5. Chattopadhyay *et al.* (1988) suggested 10-20ppm BOD_5 load of pond water to be optimum for aquaculture. The average BOD_5 level in experimental pond was observed as 6.15±3.64 mg/l. maximum concentration level (11.5-10.5 mg/l) was observed during the month of February and September. There is a 64 per cent reduction in BOD level of experimental pond than the control pond, which was due to partial digestion of piggery effluent by azolla in waste stabilization pond.

Phytoplankton are chief primary producers in fish ponds. They constitute main food for zooplankton and majority of fishes. The three most common groups of algae in fresh water fish ponds are chlorophyceae, cynophyceae and bacillariophyceae. The production of the basic food by these algae in the process of the photosynthesis is called primary production which can be measured as gross or net primary productivity. The primary productivity of phytoplankton depends mainly on the factors such as temperature, nutrient concentration and their ratio. In present experiment, phytoplankton community was constituted by 18 members with dominance of the members of chlorophyceae and bacillariophyceae. Singh and Kumari (1994) also reported Chlorophyceae as largest group of phytoplankton. Species composition, density of phytoplankton and seasonal changes in their population are determined by interaction between various abiotic and biotic factors. Total phytoplankton density (210 nos./l) was low during the winter season when the temperature and intensity of light were low indicating the preference of majority of the algal species for the higher temperature and light intensity. During the period of higher phytoplankton density, concentration of free CO_2 was low due to greater utilization by phytoplankton in photosynthesis (Mukherjee and Pankahakshi, 1995). Nutrient status of water bodies also greatly influences the diversity and density of phytoplankton (Westlake, 1980). Number of species were higher during winter. During the highest production of phytoplankton biomass, available nutrients (NO_3 and PO_4) were relatively low as utilized by phytoplankton. Majority of the species of chlorophyceae except *Spirogyra* sp. and *Oedogonium* sp. exhibited similar pattern of fluctuations being in higher numbers during April-May and July August, However numerical density of spirogyra and *Oedogonium* sp. was higher during the winter month. A number of studies in India have also reported the similar behavior of above chlorophycean species (Phillipose, 1959; Sharma, 1980).

Primary productivity, which is considered to be a good measure of phytoplankton abundance and activity in aquatic ecosystems are generally influenced by the extent of availability of solar radiation, temperature and concentration of essential nutrients. Wetzel (1966) has observed that high rate of primary production was almost always associated with an increased availability of solar radiation. Goldman and Wetzel (1963) and Efford (1967) have established a close relationship between light penetration in a water body and its productivity. Rao *et al.* (1982) have noticed positive correlation of GPP with dissolved oxygen and pH and negative correlation with free CO_2. In present experimental ponds also, the rate of energy fixation by phytoplankton showed considerable variations owing to seasonal changes in water quality parameters. It was observed that the periods of highest GPP in the experimental pond corresponded with highest dissolved oxygen, high pH and minimum carbon dioxide.

Observations of Sreenivasan (1976) have revealed that temperature was another important factor regulating primary production. Rao *et al.* (1982) established a positive correlationship of GPP with temperature. The highest GPP values recorded in the experimental ponds during April and May, therefore, might be due to the higher water temperature during these months.

Primary productivity mainly depends on the amount of nutrient supplied to the pond. Controlled application of organic manures and fertilizers in fish ponds increases the level of primary production (Mitra *et al.*, 1987; Ovie and Adenji, 1991). Several workers have observed considerably higher values of GPP in ponds fertilized with the manures than the unfertilized ponds. Highest mean value of GPP (0.580±0.3 gC/m^3/hr) was observed in the experimental pond, which received highest quantities of nutrients in the form of organic manure and atmospheric fixed N_2, while control pond in which relatively less quantity of nutrients was supplied, had registered lower mean value (0.52±0.26 gC/m^3/hr) of GPP.

Observations of Wright (1959) indicated that phosphate was a limiting factor for photosynthesis. Efford (1967), Khan and Siddiqui (1971), Sharma (1980) and Pant *et al.* (1981) have also reported close relationship between nutrients and phytoplankton primary production. Consequently, the higher concentrations of phosphate and nitrate during April and May might have been responsible for higher values of GPP in the experimental pond.

The values of GPP in present experimental ponds ranging between 0.208-1.240 gC/m^3/hr (5800 mg C/m^3/day) were comparable to the values of GPP (1009-8500 mg C/m^3/day) recorded by other workers in fertilized ponds (Hepher, 1962; Boyd, 1973; Rao and Raju, 1989, Ayyappan *et al.*, 1990).

Rodhe (1969) has classified the fresh water bodies having an annual gross primary production of more than 350 g C/m^3/yr as polluted ones. On the other hand, Vollenweider (1968) has suggested that GPP values in eutrophic water bodies could be several times higher than this value. Thus an average GPP value 2117 g C/m^3/yr recorded in the present experiment exhibit the nutrient enriched status of these ponds. The mean values of community respiration in the experimental pond found 48.28 per cent of GPP. Higher values of community respiration in experimental pond and control pond could be due to the higher standing biomass of plankton and bacteria in these ponds, which increases the rate of respiration than the production. Ayyappan *et al.* (1990) recorded community respiration values of 47-124 per cent of GPP. Ganapati and Sreenivasan (1972) have also reported high values of community respiration in some reservoirs and lakes. Ganf (1972, 1974) has concluded that higher values of community respiration (40 per cent of GPP) are the characteristic feature of eutrophic water. Thus, higher values of community respiration in present experimental ponds also indicate the nutrient rich conditions of the ponds.

The present results revealed that higher amount of organic nutrients supplied in experimental pond could sustained higher phytoplankton density and primary production.

The population of zooplankton is largely dependent on the phytoplapton population. Majority of rotifers, cladocerans and copepods feed on phytoplankton. Besides phytoplankton, the zooplankton invariably utilized detrital organic particles. Some species of copepods and rotifers are predatory in nature. A total of 7 species of zooplankton were observed with abundance of rotifers in December and occurrence of cladocerans in all months. The species diversity of zooplankton is reported to be higher in temperate regions compared to the tropical climate (Fernando, 1980; Dumont and Tundisi, 1984 and Vijverberg *et al.*, 1987). The number of species registered in a sub temperate water body was considerably higher (Sharma, 1980) than the tropical and sub tropical water bodies (Ruttner, 1952; Lewis, 1979; Singh *et al.*, 1990 and Rawat, 1991). The zooplankton communities in the present experimental ponds are reflected the tropical ecosystem with dominance of only few species. Further, relatively low species richness in the present experimental ponds may be due to higher concentrations of the nutrients in these ponds which tend to reduce the species richness. Rotifers were more dominant (61 per cent) in total composition of zooplankton population of control pond. Rotifers are well known indicators of eutrophication (Ali *et al.*, 1990). Among cladocera, *Daphnia* spp. and *Moina* sp. were abundant through out the year in both ponds, while there density was higher in experimental pond. Abundance of majority of cladoceran species indicates a relatively cosmopolitan nature of most of these species (Davis, 1972). Copepods in the experimental pond were represented by two species *viz. Cyclops* spp. and *Diaptoms* spp., between them *Cyclops* was extremely dominated. Two peaks of these species were noticed during winger and summer. Abundance of these species in tropical and temperate water was also reported by Pennak, 1957; Sharma and Pant, 1984 and Datta *et al.* 1987. Though, the total count of zooplankton (160-636 nos./l) indicates good productivity as it was reported in the range of 27-558 no/l by Singh and Sharma (1998). Two peaks were observed during January and April. Wetzil (1975) also reported two peaks of zooplankton production. Seasonal variations in the population was greatly affected by the variations in the phytoplankton population. Total zooplankton population had highly positive correlation with density of phytoplankton and GPP. It appears that multiplication rate of phytoplankton was more than the grazing rate by zooplankton. Volume of the plankton was observed in the range of 0.6-4.2 ml/50 lit. (mean 2.16±1.07 ml/50 litre) in experimental pond and 0.4-3.6 ml/50 lit. (mean 1.97±0.95 ml/50 litre) in control pond, which reflects the good production of natural fish food organism and superiority of experimental pond productivity over control pond.

Benthic fauna constitutes the important source of live food for fishes. Benthos are mainly detritophagus in nature and constitute very important component in fish food chains. The benthic organisms also create bioturbation during their movement and feeding activities, conditioning the sediments and acting as stimulants for nutrients regeneration. Study of seasonal changes and distribution of benthic fauna is very helpful in predicting the productivity of the water body. Benthos recorded in experimental pond were mainly represented chironomids, oligochaets and molluscs. Among them chironomids were prominent group with population varing from 827-2680 nos./m^2 followed by molluscs (93-468 nos./m^2). Singh and Singh (1996) also reported the dominance of dipterans. Density and biomass of chironomids were lowest during early part of the study and increased gradually with the progress of time. The rate of increase was more rapid after March attaining maximum values during May.

Mandal and Moitra (1975) also recorded maximum density of benthic fauna during early summer. Abundance of all the benthic faunal groups during March-May may be attributed to the increase in temperature with increasing rate of decomposition of organic matter leading to increased availability of food to benthic animals. Presence of large amount of dipterans and other benthos in pig-fish pond indicates that the bottom is rich in nutrients. This condition has been reported for eutrophicated lakes where chironomids is considered as an indicator of high tropic status (Gupta and Pant, 1983) Comparatively high density of molluscs in experimental pond with highly significant difference may be due to azolla integration, which facilitates the production of snails.

Qualitative and quantitative availability of fish food organisms, which depends upon availability of important nutrients in pond soil and water, determines the density of fish to be stocked in the pond and governs the rate of fish growth. Manures, fertilizers and other organic matter are the major source of nutrients, which are made available to fish food organisms, after their microbial decomposition in pond. The low concentration of nutrients results into poor pond productivity, where as excessive concentration of nutrients leads to the deterioration of water quality. Therefore, balanced concentration of nutrients is regulated by manipulating the dose of organic inputs. The effect of differential amount of organic nutrients added in experimental pond, on the fish production has been clearly reflected in the present experiment.

Fish grew at faster rate in early life stages. Similar observation noticed by Singh *et al.* (1972). Low temperature prevailing during winter season reduces the food consumption by fishes, hence lowering the growth.

Although, the average survival rate of all fishes pooled in both ponds was in the range of 78.65-98.95 per cent, which was lower as reported earlier (85-100 per cent) by Borah *et al.* (1998). The important factor responsible for lower survival rate was higher temperature in summer months and lower temperature in winter, the climatic condition of the area. Higher mortality rate of surface feeder *Catla catla* (21.35 per cent) than the other species supports, the effect of this causing factor for higher mortality. Singh (2002) reported the survival of these species in the range 67-72 per cent in an integrated fish production practice. In pig fish integrated pond, Singh *et al.* (1972) obtained survival rate of fishes in the range of 80.0-98.9 per cent of fishes with best survival of surface feeder fishes. The survival rates in the experimental pond of present study (81.25-98.95 per cent) are comparable to the above values. The surface feeder, *Catla catla* contributed highest share (40.53 per cent) to the total production which compares favorably with the excellent growth of surface feeder recorded by Sharma *et al.* (1983). Borah *et al.* (1998) obtained an average net weight as 700g for *Catla catla*, 835g for *Labeo rohita*, 740g for *Cirrhinus mrigala* and 625g for *Cyprinus carpio* in a trial of 11 months under poultry-pig-fish integrated system. The average net weight in the experimental pond of present study was recorded as 844.2g for *Catla catla*, 692.2g for *Labeo rohita*, 680.9g for *Cirrhinus mrigala* and 1045.5g for *Cyprinus carpio*. Abundance of the plankton may be the reason of higher growth of *Catla catla*, but lower average net weight of *Labeo rohita* may be due to lack of macrophytes in the pond as these fishes feed on decayed fragments of aquatic vegetation. Abundance of benthic fauna and detritus substances has provided favorable growth conditions for *Cyprinus carpio*. Comparatively low growth of *Cirrhinus mrigala* may be due to feeding compatibility with *Cyprinus carpio* as both are bottom feeder fish. A total of 20.06 per cent increase in total yield of fish reflects the superiority of pond productivity and better survival in azolla-pig-fish integrated practice over the existing pig-fish integrated system. Out of which 4 per cent was contributed by better survival rate which may be due to prior digestion of waste by azolla and 16.06 per cent was contributed by better pond productivity, which may be due to rich nutrient status in experiment pond. The abundance of green and blue green algae may be the prime reason for enhanced growth of these plankton phagous fishes in experimental pond. High production trend in the experimental (5315 kg/ha) and control (4427 kg/ha) pond found to be similar with the earlier reports on pig fish farming in India and abroad (Sharma *et al.*, 1979; Cruz and Shehadeh, 1980; Woynarovich, 1980; Sharma and Das, 1988). Such high rate of fish yield was due the application of the animal excreta, recycled in the pond, which served two most important purposes for enhancing fish yield (as direct feed and pond fertilizer) and also acted as substratum for multiplication of microbial community that provide essential nutrition for fish and fish food organisms (Newell, 1980; Schroedar, 1980). An uniform linear growth pattern was exhibited by almost all species in experimental pond. The over all SGR showed in the range of 0.6246-0.7611 by fishes pooled in experimental pond which is almost similar as reported earlier by Kumar (2004). The growth increment pattern exhibited in experimental pond found to be highly significant on statistical evaluation by subjecting students 't' test at 0.05 level. Increasing growth pattern and high yield of fish shows the sustainability of azolla-pig-fish farming system over the existing pig-fish culture practice. Overall high protein content (17.99±0.12 per cent) in all fishes pooled in experimental pond indicates the better protein utilization by fishes under azolla-pig-fish culture system. It was quite different in *Catla catla* as well as in other species also. Lower fat content (1.48±0.33 per cent) and higher ash content (6.75±0.42 per cent) also show better nutritive value of fishes, grown in experimental pond.

The values of regression coefficient 'n' (2.7121 to 2.9135) are in agreement with the studies of Quadiri and Mir (1980), Chauhan and Malhotra (1981), Malhotra (1982, 1985) and Malhotra and Chauhan (1984).

Thakur and Das (1974) stated that the value of an exponent significantly greater than 3 or less than 3 denoted that it did not maintain the isometric pattern of growth. This means that if the exponent is less than 3, the species becomes lighter for its length as it grows larger and if greater than 3, the species becomes heavier for its length at it grows longer. In the present study, these values are very close to 3.0 in fishes of experimental pond than in the fishes of control pond, indicate the more isometric growth of fishes of experimental pond. Thus the results of the present study are in conformity with the views of Le Cren (1951) and Chauhan (1987) that as the fishes normally do not retain the same shape or body outline throughout their life-span and specific gravity of tissues may not retain constant, the actual relationship, may depart significantly from the cube law. Various workers have calculated the ponderal index or condition factor of different fishes *viz.* 0.73 to 0.95 in *Tor putitora* (Pathani and Das, 1980) 1.03-1.31 in *Salmo trutta fario* (Kumar *et al.*, 1979). 1.20-1.3 in salvelinus namaycush (Oosten and Eschnerger, 1956), 1.17-1.26 in *Labeo rohita* (Kumar, 2004). Kumar *et al.*, 1979 concluded that a fish recording the value of condition factor as about one is considered to be of its average weight. This factor is the indicator of the robustness or well being of fish. The results of the 'k' values of different species in experimental pond (1.1515-1.6906) are in conformity of above results, which shows the well being of fishes in the experimental pond.

Live stock raising is apparently a cost intensive and low profit oriented practice primarily due to involvement of high cost of animal feed. In present study it was also found that the nitrogen content of the pig waste of pigs fed with azolla supplemented diet is high (1.88

per cent) than the nitrogen content in waste of pigs fed with non azolla diet (1.78 per cent). The production of pigs was almost similarin both the sets of experiment, which revealed that substitution of 20 per cent of standard feed by azolla biomass did not hamper the health and growth of pig significantly in the experimental unit. Azolla is quite efficient to fulfill the protein requirement of pig, as contain 22 per cent crude protein. Further in the present study the cost of production of pig was reduced to 14.5 per cent through utilization of azolla in pig diet. In present investigation, an average gain in live weight of pigs was recorded as 63.05 kg and 64.45 kg with azolla supplemented diet and 61.95 kg and 62.30 kg with non azolla fed diet in six months. These results are in the coincide that azolla supplemented feed contain higher crude protein level (17.2 per cent) than non azolla commercial feed (16.0 per cent). Maricel *et al.* (1990) reported 0.363 and 0.341 kg daily gain in live weight of pigs fed with 15 per cent and 30 per cent level protein replacement by azolla in growing phase of pig rearing. For finishing phase, it was 0.589 and 0.568 kg with 15 per cent and 30 per cent level protein replacement. In present study 0.358 kg daily gain in live weight of pigs was observed, with 20 per cent protein replacement by azolla, which is almost similar to the above report.

One of the principal limiting factor to profitability of a pig enterprise is the cost of the protein supplement. The expensive protein ingredients of the pig diet can be replaced by foliar sources of protein in the form of leaves harvested from forage crops, trees and water plants. Mena (1988) also reported that one of the problems with leaves from tropical trees is the presence of secondary plant compounds, chiefly phenolic substances which make the leaf unpalatable especially for monogastric animals.

Lumpkin and Plucknett (1982) and Van Hove and Lopez (1983) reported that the aquatic plant species, because of their growth habit appear not to accumulate secondary plant compounds and therefore offer perhaps a greater potential than trees as a source of protein for manogastric animals. Water fern azolla filiculoides, which grown in association with the blue green algae *Anabaena azollae* is perhaps the most promising from the point of view of cultivation, productivity and nutritive value. They pointed out that the limiting factor to the use of fresh azolla in the diet of pigs is its low nutrient density due to the extremely high moisture content (approximately 90 per cent) and the inability of the young pigs to consume the required quantities. How ever, they recommended that during the growing phase of piglets, it is necessary to offer the azolla in two portion daily while in the finishing period it is sufficient to feed it only once in morning. In the present study, slight difference was observed in the average feed conversion ratio between group of pigs fed on azolla supplemented diet (95.62 in first lot and 5.79 in second lot) and group of pigs fed on commercial without azolla supplemented diet (5.76 for first lot and 5.92 for feed second lot). Maricel *et al.* (1990) also reported beatter FCR of azolla fed pigs in finishing phase. In present study, 2.49 per cent increase was observed in total live weight of pigs with 20 per cent level supplementation of sun dried azolla in commercial diet.

After the study on four different variety it was concluded that *A. pinnata* asiatica and *A. macrophylla* is suitable to grow in north Indian conditions. These results are in conformity of Sounders and Flower (1992). Brown raddish coloration was observed during winter season. Sudden death of azolla biomass was observed during summer, when the temperature was above 38°C. similar results have been reported by Lumpkin (1977). Shade over the pool and lowering down water level (30 cm) from ground level of azolla pool was experienced as good protective measure against these adverse conditions. Low water level (20-30 cm) and low pH of water (4.5-6.0) were also experienced for better growth and short double time of azolla. Ashton (1974) also observed the fastest growth of azolla in the pH range 4.5-7.0. An 1.74 kg/m^2/week net production was recorded with 3 days of doubling time except extreme cold (<20°C) and hot (>38°C) days. Gavina, 1994 and Maricel, 1990 have also reported 3 days doubling time. Proximate composition of azolla biomass recorded as 94.2 per cent moisture, 22 per cent crude protein, 6.1 per cent fat and 18.5 per cent crude fibre, which are in conformity of Gavina, (1994). Azolla forms a permanent, heterocyst-forming cynobacterium, *Anabaena azollae,* occurs as filament located on the plant stem apexes and inside the leaf cavities. It can develop on a medium devoid of nitrogen compounds because of the ability of reduce nitrogen to ammonia and ammonia to nitrate by bacteria, present in leaf cavities. Work on nitrogen fixation capability of azolla has been done by Espimas and Watanable (1976) who reported N_2 fixation by azolla as 0.4-0.85 mg/day/g fresh weight. In present study, it obtained as 0.88 mg/day/g fresh weight, which is very close to the above report.

Faecal matter or piggery effluent is the best source of nutrients to be recycled into fish pond. The averge faecal matter production by an individual animal was noted as 794.37 g in the present study. It was reported earlier as 389.2 g by Malik (1999). According to the proximate composition, it was noted that 246.53 kg/yr/ha nitrogen and 52.45 kg/yr/ha phosphorus is added in the pond keeping 46 pigs. It found in the present study that integration of azolla in existing pig-fish culture system increases the available nitrogen as 5 per cent and decrease phosphorus as 11 per cent from the recycled waste. Budget of nitrogen was calculated in experimental and control pond. On the basis of proximate composition of recycled of waste and azolla, biochemical composition of fish plankton and benthic community and existing level of nitrogen in pond water and sediment. It was not work out by any standard method. Pig excreta and nitrogen fixation by azolla were considered as major sources of inputs of nitrogen in experimental pond, while pig excreta as single source of nitrogen into in control pond. Marco *et al.* (1994) also reported approximately 2 per cent nitrogen supply through water supply. But, it was not considered in present study as the tube-well water had used for intermediate filling of ponds and had negligible concentration of nutrients. Briggs and Fung Smith (1994) could account only 83-91 per cent of the gross nitrogen in pond while calculating the budget, while it was 89.7 per cent in the present study. About 10.3 per cent of the gross nitrogen input could not be traced. It may be attributed to the losses through ammonia, balatization, denitrification, leaching, calatization, denitrification, leaching evaporation, seepage etc. in present study, 17.83 per cent of the total amount of nitrogen was present in sediments. Briggs and Furge Smith (1994) and Martin (1998) recorded 31 per cent and 38.4 per cent nitrogen balance in sediments, respectively. Nitrogen values incorporated in fishes (52.92 per cent) in the experimental pond were comparable to those reported by Boyd (1985), Krom *et al.* (1985), Macro *et al.* (1994) and Martin (1998). Contribution of azolla in total nitrogen addition into pond is 17.2 per cent in the form of atmospheric nitrogen fixation, which is higher than the utilization of N_2 in azolla production (12.2 per cent).

Raw pig effluent is offensive in terms of its odour and also have high BOD, which is the amount of oxygen required for bacterial decomposition. Treatment of the effluent can be done in an aerobic digester in the form of aerobic lagoons. Aerobic system can be naturally or mechanically operated. In natural systems, oxygen is aided by growth of algae which utilize waste as source of nitrogen and the sun energy. It has now been demonstrated that aquatic macrophyte-based water treatment system (AMS) offers a low energy consuming and low cost method from removing contaminants (Rengaraj *et al.*, 1999). Plants cultured over waste water perform several functions including assimilating, storing contaminants transporting oxygen and providing substrate for microbial activity which helps in the stabilization of wastes (Rengaraj *et al.*, 1999). Ready and DeBusk (1985) reported the remarkable ability of aquatic macrophytes in transporting atmospheric oxygen to waste water. Flessa, 1994, reported that a portion of transported oxygen by aquatic macrophytes is consumed during root respiration, the rest enters into the water column and is utilized by facultative and aerobic bacteria. In present study, azolla grew on the surface of waste stabilization pond. More or less facultative condition were prevailed in this pond and decomposition of organic matter was done by aerobic and facultative bacteria. One week old semi digested piggery waste was used to recycle into experimental pond, which resulted as low BOD level of pond water and readily availability of nutrients, altimately improved the fertility of experimental pond.

The actual cost and return of poultry-pig-fish farming and pig-fish farming was worked out by Borah *et al.* 1988. He reported 40.5 per cent reduction in total variable cost, with 87 per cent return in poultry pig-fish farming. The net return of the experimental unit was Rs. 20375.50 with 77.66 percent return in variable cost and it was Rs. 13101.55 with 44.74 percent return in variable cost of control unit. The net economic return was 55.52 per cent higher in the azolla-pig-fish culture in comarision to pig-fish culture and was found to be significant on subjecting to student's 't' test at 0.05 level. Gavina (1994) reported integration of azolla with duck-fish integrated farming system. Present study reveals the intigration of azolla-pig-fish farming. Better pond productivity was experienced in this integrated system due to more input supply of nutrients into pond, capability of azolla for atmospheric N_2 fixation and semi-digestion of piggery waste by azolla. Better productivity, more production of fish as well as pigs, reduction in pig production cost and high per cent return over the variable cost (77.66) reflects the technical and economical feasibility of this integrated model. The finding of the study reveals the efficacy of the system.

Chpater 7

Summary

The present thesis "Integrated Azolla-Pig-Fish farming with special reference to pond productivity and fish production" mainly deals with physio graphical features and meteorological parameters of study area, impact assessment of recycling of piggery waste on chemical properties of pond soil, physico-chemical and biological parameters of pond water specially emphasizing pond productivity and fish production with economics of integrated unit. The main findings of these investigations are summarized as follows:

1. Experiment was conducted for 12 months duration at Fish Production and Pig Breeding Farm of Krishi Vigyan Kendra, Rampura, Rewari (Haryana), which is located in the south western part of Haryana State and lies between 27.5°- 28.5°N latitudes and 76°- 76.5°E longitudes, near Rewari City. This area is affected by semi-arid climate of Rajasthan.
2. Experiment was conducted in two earthen ponds of same size (0.12 ha) situated very close to piggery unit. In experimental unit, fish pond was integrated with pig sty and azolla production-cum-stabilization pond (6×4×0.6m) while control unit was designed only to integrate fish pond and pig sty. Fish ponds were stocked with advance sized fingerlings of *Catla Catla, Labeo rohita, Cirhinus Mrigala* and *Cyprinus carpio* Var. Communis @8000 fingerlings/ha in

the ratio of 4:3:2:1. Weaner piglets of the variety Large White Yorkshire of 2½ months age were stocked in each pen in 6 numbers for first 6 months and 5 numbers for second six months. *Azolla pinnata* was grown in waste stabilization pond.

3. All sub units of each unit were connected with PVC pipeline. One group of pigs was fed with commercial diet and waste was recycled in control pond, another group was fed with 20 per cent azolla supplemented commercial feed and waste was digested in stabilization pond, then released into experimental pond on weekly interval. Azolla was produced in stabilization pond and used for feeding of experimental pigs.
4. Wide fluctuation was experienced in ambient temperature (3.5-48.5°C) with very less (88.2 mm) rainfall during study period.
5. Slightly alkaline (7.4-7.6) pH, 1.22 ± 0.26 per cent organic carbon with slightly elevated concentration during summer months, 49.14 ±10.48 mg/100g soil available nitrogen with gradually increasing trend, 4.94 ± 1.1 mg/100g soil available phosphorus with significant difference from control unit was observed for pond sediment. All parameters except soil pH showed significantly positive correlation with GPP.
6. Wide seasonal fluctuation was observed with almost similar values of physico-chemical parameters in both ponds except for dissolved oxygen, nitrite-nitrogen and BOD. Temperature of pond water found in the range of 14.7-32°C. pH values varied from 6.9-8.0 (7.54 ± 0.33). Dissolved oxygen level was observed in sub optimal level (5.41±0.77 mg/l) with lower level (4.04 mg/l) during winter months. Free CO_2 was found in the range of 0.0-9.2 (5.72±2.75) mg/l with lower concentration during summer and higher during winter. Fluctuation in total alkalinity found in the range of 91-177 (155.3 ± 24.7) mg/l. Salinity of the pond water was below critical level (0.4-2.2 ppt) with higher concentration during summer months. Continuous fluctuation found in the values of TDS (180-298 mg/l).

7. Concentration of ammonia-nitrogen, nitrite-nitrogen and nitrate-nitrogen were observed in the rage of 0.01-0.24 mg/l, 0.017-0.055 mg/l and 0.07-1.22 mg/l, respectively. All types of nitrogen were minimum during winter season and maximum during summer except NH_3-N, which was maximum during spring season. Bimodal pattern of fluctuation was observed in phosphorus content with mean value 0.23±0.08 mg/l Remarkable difference was observed in the values of BOD_5 with a level of 16.85 ± 3.85 mg/l in control pond and 6.15±3.65 in experimental pond. All physico-chemical parameters of pond water are positively correlated with GPP except pH, free CO_2 and BOD_5.
8. Total phytoplankton density was lower during winter with mean value 1347.83±1010.64 nos./l and it was higher in the month of April with average plankton volume (2.16 ml/50 l). Fluctuation was observed due to seasonal variation in temperature, free CO_2 and nutrient status. A total of 18 genera of phytoplankton were recorded with maximum composition of green algae (60.3 per cent) followed by diatoms. Dominance was occurred by green algae in rainy and summer season, by diatoms and dinoflagellates in rainy season, by blue green algae in winter and by Euglenophytes in winter and summer season. Among green algae, occurance of *Spirogyra* and *Oedogonium* was higher in winter season.
9. A total of 7 species of zooplankton were observed with abundance of rotifers in December and occurance of Cledoceran in all months. Total density varied from 27-588 nos./l with almost equal occurrence of rotifers and Cladocerans and dominance of *Dhaphnia* spp and *Moina* spp. Two peaks were observed during January and April with seasonal variation in population due to phytoplankton population.
10. Gross primary productivity (GPP), Net primary productivity (NPP) and Community Respiration (CR) showed wide seasonal fluctuation with mean values 0.58±0.3, 0.31±0.18 and 0.28±0.16 gC/m^3/hr, respectively. Period of highest GPP values (April-May) coinsides with

the highest NPP values. Low values of CR were observed in initial of experiment and in winter months. GPP showed significantly positive correlation with all physico-chemical parameters of pond water and chemical parameters of pond soil except water pH, free CO_2, BOD and soil pH.

11. Among benthic fauna diptera represented by *Chironomus* spp with varied range of 827-2680 (1803±626.85) nos./m^2, olegochaets represented by *Tubifex* spp and *Dero* spp with range of 65-390 (208.83±98.27) nos./m^2 and Molluscs represented by *Gyraulus* spp, *Limnaea* spp and *Lamellidens* spp with significant difference from control pond (93-263 nos/m^2). Gradually increasing trend was observed in the population of benthic fauna.
12. The average survival rate of fishes was 86.63 per cent with highest for *Cyprinus carpio* (98.95 per cent) and lowest for *Catla catla* (81.25 per cent). Average net weight was highest for *Cyprinus carpio* (1045.5 g) followed by *Catla catla* (844.2 g). A net fish production of 637.8 kg (5315 kg/ha/yr), was obtained from azolla-pig-fish farming, while it was 531.2 kg (4427 kg/ha/yr) in pig-fish farming. Out of 20.06 per cent increase in total production, 16.06 per cent was due to better productivity and 4 per cent due to better survival. An uniform linear growth pattern was exhibited by almost all species with SGR in the range of 0.6246-0.7611. The values of regression coefficient 'n' were in the range of 2.7121 to 2.9135 with condition factor 'k' in the range of 1.1515-1.606, showed isometric pattern of growth and better favourable conditions in experimental pond. Over all high protein content in all fishes (17.99±0.12 per cent) indicates better protein utilization by fishes and lower fat content (1.48±0.33 per cent) and higher ash content (6.75±0.42) showed better nutritive value of experimental pond grown fish.
13. Azolla is quite efficient to fulfill the protein requirement of growing pigs, having 22 per cent crude protein in proximate composition. A reduction of 14.5 per cent was noted in total pig production cost of experimental unit

with 20 per cent supplementation of sun dried azolla in pig diet. Azolla supplemented feed showed higher protein level (17.2 per cent) than the common pig feed (16.0 per cent). Proximate composition of pig dung voided by azolla (20 per cent level) fed pigs, recorded as 1.88 per cent nitrogen, 0.6 per cent phosphorus, 0.33 per cent potassium and 70 per cent moisture content. A total yield of 845 kg pigs was obtained with average live weight 63.05 and 64.45 kg in first and second lot, respectively. Hence, more nutrients in pig dung and 2.5 per cent more production of pigs with 14.5 per cent reduction in cost of pig production was noted after supplementation of azolla in pig diet.

14. Among the four tested varieties of azolla, satisfactory growth was noted for *Azolla pinnata* and *Azolla microphylla*. Brown radish colouration and reduced growth found during winter season (below 20°C). Optimum temperature for proper growth of azolla found in the range of 20-30°C and pH in the range of 4.5-7. An 1.74 kg/m^2/week net production was obtained with 3 days of doubling time during climatic conditions. Nitrogen fixation rate worked out as 0.88 mg/day/g fresh biomass. Proximate composition of azolla recorded as 94.2 per cent moisture, 22 per cent crude protein 6.1 per cent crude fat and 18.5 per cent crude fibre. It was worked out that 9 kg/week azolla biomass is required to supplement 20 per cent diet of a pig, which can be obtained from 5m^2 cultivation area.
15. A total quantity of 1513.55 kg piggery waste was recycled into pond with an input supply of 146.53 kg/ha/yr nitrogen and 52.42 kg/ha/yr phosphorus. Contribution of azolla in total nitrogen addition in the form of N_2 fixation was higher (17.2 per cent) than the utilization rate (12.2 per cent), showed 5 per cent increase in total nitrogen budget. Allocation of total nitrogen budget worked out as 3.6 per cent in pond water, 17.8 per cent in pond sediment, 1.9 per cent in plankton, 1.2 per cent in benthos, 52.9 per cent in fishes, 12.3 per cent azolla biomas and 10.3 per cent as untraced.
16. Azolla not only served as feed ingredient but also performed macrophyte based treatment of piggery effluent

in stabilization pond and resulted as 64 per cent reduction in BOD level of experimental pond.

17. In total cost of production (Rs. 29282), 10.4 per cent reduction was noted due to supplementation of azolla in pig diet. The net return of the experimental unit was Rs. 20375.50 with 77.66 percent return in variable cost. The net income return was 55.52 per cent higher which was due to reduction in production cost and more production of fish and pigs.
18. The findings of the study reveals the better pond productivity with rich nutrient status, better fish survival, more production of fish and pigs, efficacy of azolla to fulfill protein requirement of pigs as feed ingredient as well as potentiality to digest piggery waste in stabilization pond and better economic return. On the basis of these findings, it can inferred that integrated azolla-pig-fish farming is highly sustainable aqua farming system.

References

Abbasi, S.A. 1987. Aquatic plant-based water treatment systems in Asia. K.R. Reddy and D.W.H. Smith (Eds.). Aquatic plants for water treatment and resource recovery. Magnolia Publishing Inc.; Orlando. 175-98.

Abbasi, S.A. Baji, V. and Kunhahmed, T. 1988. A hydroponic system for studying the impact of water quality on the germination and early growth of angiosperms. *Environmental Technology.* 9: 219-226.

Ahmad, S.H. and Singh, A.K. 1991. Diurnal rhythm of zooplankters and their correlation in a fresh water pond of Dholi, Bihar. *Environment and Ecology,* 9(1): 23-28.

Akpan, E. and Okafor, N. 1997. On organic fertilization and plankton development in two experimental fresh water ponds of Nigeria. *J. Aqua. Trop.* 12: 147-154.

Alabaster, J. S. and R. Lloyd, 1980. Water quality criteria for fresh water fish. FAO publication. pp. 297.

Alam, W.; Sheri, A.N. and Afzal, M. 1990. Effect of poultry manure on the growth performance of *Labeo rohita* under polyculture system. *In: Proc. Pakistan Congress of Zoology.* 10: 301-306.

Alejar, A.S.; Mabbayad, B.B. and Escobin, R.P. 1984. Rice-duck integrated farming system Mimeographed progress report.

Ali, M. Khan, A.A. and Haque, N. 1990. Proc. Nat. Acad. Sci. India. 60(B) 1: 13.

Ali, S. A. and Newton, S. 1988. Prospectus of fish culture in waste water system. Aqua. Mag. Philadelphia, USA.

Alikunhi, K.H. 1957. *Fish culture in India. FM. Bull.* Delhi, ICAR. 20: p. 144.

Anon. 1985. Package of practices for animal husbandry and dairying. Assam Agricultural University, 1985.

Anon. 1993. Package of practices for fish farming in Assam. Composite culture of carps and barbs, Assam Agricultural University 1993.

Anon. 1997. *Package of practices for fish farming in Assam.* A.K. Bhagbati (ed.) Assam Agric. Univ., Zorhat. 49: p. 52.

Antoine, T.; Wery, P.; Micha, J.C. and Van Hove, C. 1997. Comparison of the growth and chemical composition of *oreochromis (Tilapia) niloticus* and *Cichlassoma (Theraps) Melanurum* fed with Azolla. *Aquaculture,* 66(1): 81-96.

Anuta, J.D. 1995. Effect of chicken manure application on water quality and production of Tilapia guinensip in fresh water concrete tanks. *J. of Aqua. In the Tropics.* 10(2): 167-176.

AOAC 1980. Official methods of analysis of the association of analytical chemist (Ed. W. Horwitz), 13th edition, Washington, D.C., p. 1094.

APHA 1985. *Standard methods for the examination of water and waste water,* 16th ed. American Public. Health Association, New York, p. 1268.

APHA. 1971. Standard Methods for the Examination of Water and Wastewater Washington, D.C. pp. 874.

Arce, R.G. and Boyd, C.E. 1980. Water chemistry of Alabama ponds. Auburn University Agricultural Experiment Station, Auburn University, Alabama, Bulletin, 522, p.35.

Armstrong, D.A. 1979. Nitrogen toxicity to crustacea and aspects of its dynamic in culture system. 2nd *Biennial Crustacean Health Workshop, Texas.* 329-360.

Ashton, P.J. 1974. The effect of some environmental factors on the growth of *Azolla filiculoides* Lam. In: *The River Progress Report* (ed. E.M.V. Zidereen-Bakker). Institute of Environmental Sciences, University of the Orange Free State, Bloemfontein, South Africa. pp. 23-38.

Asthon, P. J. and Walmsley, R. D. 1976. The aquatic ferm Azolla and its Anabaena Symbiont. *Endeavour*. 35 (124): 39-43.

Aswathanarayana, C. 1979. Studies on the relative merits of three commonly available organic manure for fish culture. M.F.Sc. thesis, University of Agricultural Sciences, Bangalore.

Avnimelech, Y. and Lacher, 1979. A tentative nutrient balance for intensive fish ponds. *Bamidgeh*. 31(1): 3-8.

Ayyappan, S.; Pandey, B.K; Pant, K.C.; Dash, D. and Tripathi, S.D. 1991. Azolla–a nitrogenous, biofertilizer for fish ponds vis-à-vis traditional manurs. *In:* proceedings of National Symposium on new Horizons in Fresh Water Aquaculture. 23-25 January, pp. 151-154.

Ayyappan, S.; Rao, N.G.S.; Rao, G.R.M.; Janakiram, K.; Purushothaman, C.S.; Saha, P.K.; Pani, K.C.; Mudduli, H.K.; Sinha, V.R.P. and Tripathi, S.D. 1990. Production efficiencies of carp culture ponds under different management practices. *J. Aqua. Trop*. 5: 69-75.

Banerjea, S.M. 1967. Water quality and soil conditions of fish ponds in some states of India in relation to fish production. *Ind. J. Fish.* 14(1&2): 114-115.

Banerjea, S.M. and Ghosh, S.R. 1970. Studies on the correlation between soil reaction and different forms of bound phosphorus in soil. *J. Inland Fish. Soc. India*. 2: 113-120.

Banerjee, R.K.; Ray, P.; Singh, G.S. and Dutta, B.R. 1979. Poultry droppings its manorial potentiality in aquaculture. *Journal of Inland Fisheries Society of India*. 11(1): 94-108.

Barash, H. and Schroeder, G.L. 1983. Use of fermented cow manure as a feed substitute for fish polyculture in stagnant water ponds. *Aquaculture*. 36: 127-140.

Baruah, B.K.; Talukdar, S. and Borthakur, C.R. 1998. Limmnological studies of ponds in relation to pisciculture. *Environment and Ecology*. 16: 518-522.

Basudha, Ch.; Vishwanath, W. 1997. Formulated feed based on aquatic weed Azolla and fish meal for rearing medium carp osterobrama belangeri (Valencienes). *J. Aquacult.* Trop. Vol. 12 no. 3, pp. 155-164.

Becking, J.H. 1979. Environmental requirements of *Azolla* for use in tropical rice production. In *Nitrogen and Rice.* International Rice Research Institute, Manila, Philippines. pp. 345-375.

Bhattacharya, T.; Saha, R.K. and Chakraborti, J. 1988. Diel variations in the water quality, plankton population and primary production in a fresh water pond in Tripura. *Environment and Ecology*. 6(4): 928-932.

Billard, R.; Depauw, N.; Micha, J.C.; Salomoni, C. and Verreth, J. 1989. the impact of aquaculture in rural management. *Business Joins Science,* 12: 57-91.

Black, C.A. 1965. *Methods of Soil Analysis part I and II.* American Society of Agronomy, Madison.

Borah, B.C.; Bhagowati, A.K.; Baruah, U.K.; Gogai, H.K. and Nath, S. 1998. Integrated Pig-Fish farming. J. *Inland Fish Soc.* India. 30(1): 39-46.

Boyd, C.E. 1973. Summer algal communities and primary production in fish ponds. *Hydrobiol.* 41: 357-390.

Boyd, C.E. 1974. Lime Requirements of Alabama Fish Ponds. Auburn Univ. (Ala) Agri. Expt. Stn. Bull. (459). p. 20.

Boyd, C.E. 1978. Water Quality in Warmwater Fish Ponds. Agril. Expt. Stn. Auburn Univ. p. 359.

Boyd, C.E. 1979. Aluminum sulphate (alum) for precipitating clay turbidity from fish pond. *Trans. Amer. Fish* Soc. 108: 307-313.

Boyd, C.E. 1982. Water Quality Management for Pond Fish Culture. *Elsevier Science Publishing Company Inc. New York.*

Boyd, C.E. 1985. Chemical budgets for channel cat fish ponds. *Transactions of the American Fisheries Society.* 114: 291-298.

Boyd, C.E. 1989. *Water quality management and aeration in shrimp farming*. Fisheries and Allied Aquaculture Dept. Series, Aburn Univ. Alabama.

Boyd, C.E. and Pillai, V.K. 1984. Water Quality Management in Aquaculture. CMFRI. Special publication. 22: 97.

Briggs, M.R.P. and Funge-Smith, S.J. 1994. A nutrient budget of some intensive marine shrimp ponds in Thailand. *Aquacult. Fish. Manage.* 25: 789-811.

Cagauna, A. and Pullin, S.V. 1994. Azolla in Aquaculture: Past, present and furute. In: *Recent Advances in aquaculture.* (Eds. Mur, J.T. and Roberts R.J.). Blackwell Science, Combridge. pp. 104-130.

Cairns, J. Jr. 1965. Biological concept and industrial waste disposal problems. Proceeding of the 20th Industrial waste Confrence. May-4-6, Extn. Series No 118, Purdue University, Lafaytte, USA.

Chattopadhyay, G.N.; Saha, P.K.; Ghosh, A. and Karmakar, H.C. 1988. A study on optimum BOD level for fish culture in wastewater ponds. *Boil.* Wastes. 25: 79-83.

Chang, S.C. and Jackson, M.L. 1957. Fractionation of soil phosphorus. *Soil Sci.* 84: 133-144.

Chang, W.Y.B. 1987. Fish culture in China. *Fisheries.* 12(3): 11-15.

Chatterjee, D.K. and Saha, P.K. 1996. Characteristics of organic matter in undrainable fish pond soils. *Bamidgeh.* 48(3): 142-148.

Chatterjee, D.K. and Saha, P.K. 2000. Effect of organic manures on organic carbon content of submerged pond soil. *Bamidgeh.* 52(3): 118-126.

Chattopadhyay, G.N. 1998. Chemical analysis of fish pond soil and water. Daya publishing house, Delhi, p. 33.

Chauhan, D.P.S.; Sharma, L.L. and Saini, V.P. 1997. Propagating Azolla-cum-fish culture with the use of distillery effluents. *Fishing Chimes,* Vol. (17): 4-8.

Chauhan, R. S. 1987. Food parasite and length-weight relationship of a hill stream fish. *Schizothorax plagiostomus* (Heckel). *Indian J. Anim. Res.* 21(2): 93–96.

Chauhan, R. S. and Malhotra, S. K. 1981. Bionomics of hill stream cypronids II. *Barillus bola (Ham.), Biology.* 3 (2): 34-36.

Chauhan, R.S.; Sharma, A.P. and Singh, H.P. 1998. Recycling of duck excreta and its impact on fish production. *Him. J. Env. Zoo.* 12: 43-48.

Chen, D. and Huang, C. 1987. Study of Azolla as a fish fodder (Poster Paper) In: *Azolla Utilization.* Proceedings of the Workshop on Azolla use. International Rice Research Institute. Los Banos Laguna, Philippines.

Chen, T.P. and Li, Yen. 1980. Integrated Agriculture studies in Tiwan. In: R.S.V. Pullin and Zaid, H. Shedhadeh (eds.) Proc. ICLARM-SEARCA *Conf. Integ. Agri. Aqua. Fmg.* System, Manila, Philippines, pp. 239-241.

Choudhari, S. and Singh D.K. 1999. Zooplankton population of Boosra lake at Muzaffarpur, Bihar. *Envn. and Eco.* 17(2): 444-448.

CIFRI.1985. Fish cum live stock farming. Aquaculture extension manual, New series No. 5. p. 32.

CIRDAP. 1988. Report of the Regional Workshop Integrated Farming System. CIRDP Newsletter No. 34: 3-4.

Cruz, E.M. and Shehadeh, Z.H. 1980. Preliminary results of integrated pig-fish and duck-fish production tests. *In:* Pullin, R.S.V. and Shehadeh, Z.H. (eds.), *Proc. ICLARM-SEARCA Conf. Integ. Agri. Aqua. Fmg. Systems,* Manila, Philippines. 225-238.

Das, M. 1987. Effect of biogas plant effluent on aquatic biomass and fish production. Ph.D. Thesis, G.B. Pant University of Agriculture and Technology, Pantnagar, India.

Datta, N.C.; Mandal, N. and Bandopadhyay, B.K. 1987. Seasonal abundance of rotifers in a perennial fresh water pond in Calcutta. *J. Env. Biol.* 8(1): 63-71.

Davis, C.C. 1972. Plankton dynamics in a new found land lake. *Verh. Inter. Nat. Verein. Limnol.* 18: 278-283.

De Jong, J. 1976. The purification of waste water with the aid of rush or reed ponds. J. Tourbier and R.W. Pierson (Jr.) (Eds.). *Biological control of water pollution.* University of Pennsylvania Press, USA.

De Waha Baillonville, B.; Godard, T.P. and Van Hove, C. 1984. Chemical composition of Azolla populations as affected by ageing. *Arch. Int. Physiol. Biochem.* 92(1): 30.

Delmendo, M.N. 1980. A review of integrated livestock fish farming systems. *In:* Pullin, R.S.V. and Shehadeh, Z.H. (eds.), *Proc. ICLARM-SEARCA Conf. on Integ. Agri. Aqua. Fmg. Systems,* Manila, Philippines, pp. 59-71.

Dhawan, A. and Toor, H.S. 1989. Impact of organic manure on supplementary diet, on plankton production and growth and fecundity of an Indian Major Carp, *Cirrhinus mrigala* (Ham.) in fish ponds. *Biological Wastes,* 29(4): 289-297.

Dinesh, K.R.; Varghese, T.J. and Nandeesha, M.C. 1986. Effects of a combination of poultry manure and varying doses of urea on the growth and survival of cultured carps. *In: Proc. First Asian Fish Forum,* Manila, Philippines. pp. 565-568.

Dinges, R. 1976. *Water hyacinth culture for waste water treatment.* Texas department of Health resources, Division of Waste water Technology and surveillance, Austin. Texas, USA, p. 39.

Djajadiredja, R., Jangkaru, Z. and Junus, M. 1980. Freshwater aquaculture in Indonesia with special reference to small-scale agriculture farming systems in west Java, In: *Integrated Agriculture Aquaculture Farming Systems,* R.S.V. Pullin and Z.H. Shehadeh (eds), ICLARM Conf. Proc. 4, pp. 143-165.

Doudoroff, P. and Dean, L.S. 1967. Dissolved oxygen criteria for the protection of fish. Amer. Fish Soc. Spl. Pub. 4: 37.

Downing, K.M. and Merkens, J.C. 1955. The influence of dissolved oxygen concentration on the toxicity of un-ionized ammonia to rain bow trout (*Salmo gairdnerii* Richardson). *Ann. Appl. Biol.* 43: 243-246.

Dumont, H.J. and Tundisi, J.C. 1984. Epilogue, the future of tropical zooplankton studies. *Hydrobiologia.* 113: 331-333.

Dutta, N.C.; Mazumdar, A.; Ahuja, D. and Bandopadhyay, B.K. 1987. Symposium on the impact of current land use pattern and water Res. Develop. On Riverine fish. April 25-27. 1987: 45.

Dutta, O.K. and Goswami, P.K. 1988. Investigations on cow and pig manure recycling in fish ponds. *J. Inland Fish. Soc. India.* 20(1): 37-41.

Edward, P. 1985. Pigs over fish ponds. *Pig International,* 15(9) 8-10.

Edward, P. 1980. A review of recycling of organic waste into fish with emphasis on the tropics. *Aquaculture*. 21(1): 261-279.

Edward, P. 1983. The future potential of integrated farming systems in Asia.Proc. VWCP I, pp. 273-281.

Edward, P. 1986. Duck/Fish integrated farming systems. In: Duck production science and world practice, Farrell,D. J. and Stapleton, P. (eds),University of New Engl. and, U.S.A. pp. 267-291.

Edward, P., Kaewpaitoon, K., McCoy, E.W. and Chantachaeng, C (1986). Pilot small-scale crop/live-stock/fish integrated farm. AIT Research Report, 184, AIT, Bangkok, Thailand, p. 131.

Efford, *I.E.* 1967. Temporal and special differences in phytoplankton productivity in marine lake, British Colambia. *J. Fish. Res. Can.* 24: 2283-2307.

Eills, M.M. 1937. Detection and Measurement of Stream Pollution. *Bull* 22, US Bureau of Fisheries.

Elango, A.; Daniel Jameson, J. and Ramadhas, V. 1997. Production of *Chironomus* larvae for use as live food in aquaculture systems. *In:* Santhanam, R.; Ramadhar, V. and Gopalkrishnan, P. (eds.), *Proc. Nat. Sem. Water Quality Issues in Aqua. Systems.* pp. 91-101.

El-Sayad, A.F.M. 1992. Effects of substituting fish mean with *Azolla pinnata* in practical diets for fingerling and adult Nile tilapia, *Qreochomis niloticus* (L.) *Aquaculture and fisheries management* 23: 167-73.

English, P.R.; Fowler, V. R.; Boxter, S. and Smith, B.1989. Effluent storage, treatment and disposal. In: *The growing and finishing pig: Improving efficiency. Uni. of Illinois. Farming* Press, U.K. pp. 411-447.

Eren, Y.; Tsur, T. and Avnimelech, Y. 1977. Phosphorus fertilization of fish ponds in the upper Golilee. *Bamidgeh.* 29: 87-93.

Eskinazi, L.E. and Koeining, M.L. 1980. Phytoplankton composition of fish culture ponds of the region of Itamaraca. First Brazilian Symposium on Aquaculture, July, 1978: 87-97.

Espinas, C.R. and Watanabe, I. 1976. 'Potentiality of nitrogen fixing *Azolla-Anabaena* complex as fertilizer in paddy soil'. Rice Research Institute, Saturday Seminar, Manila, Philippines.

Esteky, A.A.; Olah, J.; Pekar, F.; Nazami Balouchi, S.A. and Szabo, P. 1995. Fish stresser environment in high stocked fish-cum-duck ponds. *Aquaculture*. 129: 1-4.

Fallowfield, H.J. and Martin, N.J.1988. The operation, performance and computer design of high rate algal ponds. In: Aqua. Engg. *Sympo.* Series no.111, EFCE. Pub. Series no. 66: 389-403.

Fang, Y., Guo, S.; Wang, J.; Yang, and Liu, Z. 1988. The effects of different animal manures on ecological factors and fish yields in fish ponds. In Aquaculture International Congress And Exposition Vancouver British Columbia, Canada. Sept. 6-9, pp. 48.

Fang, Y.; Guo, X.; Wang, J.; Yang, Y.; Liu, Z. and Mathias, J.A. 1994. The effect of animal manure protein on fish yield in carp polyculture. Science Press, Beijing, pp. 335-351.

Fang, Yingxue; Guo, Xianzhen; Wang, Jikun; Yejin; Yang, Yaping and Liu, Zhiyun. 1988. Effects of different animals manures on the econolical factors and fish yield in fish ponds. Network of Aquaculture Cent. In Asia, Bangkok (Thailand) 4.

FAO, 1977. Freshwater fisheries and aquaculture in China. FAO Fish. Tech. Paper No. 168, p. 84.

FAO, 1988. Integrated 'Rice-*azolla-Fish*' system, *FAO Regional Office for Asia and the Pacific.* (RAPA) *Bulletin*, 4. FAO/RAPA, Bangkok, Thailand.

Fernando, C.H. 1980. The fishery potential of man made lakes in South East Asia and some strategies for its optimization. *Biotrop. Anniversary Publ.* Boger, Indonesia, pp. 25-38.

Flessa, H. 1994. Plant-induced changes in the redox potential of rhizospheres of the submerged vascular macrophytes *Myriophyllum verticillatum* L. *Aquatic Botany*. 47: 119-129.

FOA/RAPA.1980. Report of the fifth session of IPFC Working Party on Aquaculture and Environment, Jakarta, Indonesia, 22-26 January 1980.In:Indo-Pacific Fishery Commission Proceedings, 19th session; Kyoto, Japan, 21-30 May 1980, pp. 155-164.

Fogg, C.E. 1963. The role of algae in organic production in aquatic environments. *Brit Phycol. Bull.* 2: 195-205.

Gaarder, T. and Gran, H.H. 1927. Investigations on the production of plankton in the Oslo Fjord. *Rapp. P-V. Reun. Commn. Cut. Explor. Scient. Mer. Mediterr.* 42: 1-48.

Galemoni, F. and Ngokaka, C. 1989. Evaluation of three environmental factors (O_2, Temp. pH) in an intensive culture pond at the Djoumouna fish farm. *PISCIC FR,* 98: 27-32.

Galston, A.W. 1975. The water fern-rice connection, *Natural History Magazine,* 84 (10): pp. 10-11

Ganapati, S.V. 1960. Ecology of tropical waters. *Proc. Symp. Algology, ICAR.* 204-218.

Ganf, G.G. 1972. Regulation of net primary production in lake George, Uganda, East Africa. *In:* Kajak, A. and Hillbricht- IlluKowska (eds.), *Productivity problems of fresh water.* Pol. Sci. Publ., Warsaw, pp. 693-708.

Ganpati, S.V. and Sreenivasan, 1972. Energy flow in aquatic ecosystem in India. *In:* Kazak, Z. and Hillbricht Illukuwska, A. (eds.), *Productivity problems of fresh water.* PWN, Warsaw, pp. 457-475.

Gavina, L.D. 1994. Pig-Duck-Azolla integration in La Union, Philippines: In: NAGA- ICLARM, ISSN- 0116- 290x: 18-20.

Ghosh, A.N.; Ghosh, S.R. and Sarkar, N.N, 1973. On the salinity tolerance of fry and fingerlings of Indian major carps. *J. Inland Fish. Soc. India* V, pp. 215-217.

Ghosh, S. R. and Mohanty, A. N. 1981. Observation on the effect of aeration on mineralization of organic nitrogen in fish pond soil. *Bamidgeh,* 33 (2): 50-56.

Gill, H.S. 1974. Effect of pollution upon the physico-chemical composition and fisheries of Punjab waters. Ph.D. thesis, Punjab Agricultural University, Ludhiana, India. p.147.

Golachowska, J.B. 1971. The pathways of phosphorus in lake water. *Pol. Arch. Hydrobiol.* 18: 325-345.

Goldman, C.R. and Wetzel, R.G. 1963. A study of the primary productivity of clear lake, lake county, California. *Ecology.* 44: 283-294.

Goldman, J. C. 1979. Outdoor algal mass culture, I application. *Water Research.* (13): 1-19.

Gopal, B. 1967. Contribution of *Azolla pinnata* R. Br. To the productivity of temporary ponds at Varanasi. *Trop. Ecol.* 8: 126-130.

Gopalkrishnan, P.; Ramadhas, V. and Sundararaj, V. 1997. Nitrification in a semi-intensive shrimp culture system. *In:* Santhanam, R.; Ramadhas, V. and Gopalkrishnan, P. (eds.), *Proc. Nat. Sem. Water Quality Issues in Aqua. Systems.* pp. 23-27.

Govind, D.V. Rajagopal, K.V. and Singh, G.S. 1978. Studies on the comparative efficiencies of organic manures as fish food producers. *J. Int. Fish Soc.* India. 10: 101-106.

Grower, J.H.; Renato, D. and Duraza, U.A. 1976. Production and growth of milk fish, common carp and catfish in fertilized fresh-water ponds. Kelimasan. 5: 193-206.

Guiterez, W. and Guevara, J. 1988. Preliminary results on the effects of mineral and organic (pig manure) fertilization on the production of "red tail shed" (*Brycon erythropterum*) at Pucallpa, Peru. Rev. Latinoam. *Acuiculture* 36: 35-38.

Gupta, P.K. and Pant, M.C. 1983. Macro-benthos of lake Nainital (U.P.) India with particular reference to pollution water. Air and Soil Pollution. 19: 397-405.

Gupta, P.K. and Pant, M.C. 1983. Seasonal variation in the energy content of benthic macro invertebrates of lake Nainital, U.P., India. *Hydrobiologia.* 99: 19-22.

Han, Y.G. and Ding. J.Y. 1986. Comparative study on the effects of fresh and fermented pig manure on fish yield. Bangkok–Thailand, NACA, p. 6.

Haque, N.; Khan, A.A. and Fatima, M. 1990. Composition and seasonal abundance of phytoplankton in a tropical fresh water pond at Aligarh, *Proc.* of II Fisheries Forum, Tokyo, Japan. 17 April. 1989: 361–364.

Hegde, R.N. 1981. Studies on the effect of fertilizing fish ponds with poultry manure and varying doses of NPK, M.F.Sc. thesis, University of Agricultural Sciences, Banglore.

Hepher, B. 1962. Primary production in fish ponds and its application to fertilization experiments. *Limnol. Oceanogr.* 7: 131-135.

Hickling, C.F. 1962. Fish culture. Faber and Faber, London, p. 295.

Hopkins, K.D. and Cuze, E.M. 1982. The ICLARM- CSLU integrated animal-fish farming project. Final report. ICLARM *Tech. Resp.,* 5: 96.

Hutchinson, G.E. 1957. A treatise on limnology I. Geography, physics and chemistry. J. Willey and Sons, New York and London. p. 1015.

Hutchinson, G.E. 1967. A treatise on limnology-2. Introduction to lake biology and limnoplankton. John Willey and Sons. New York.

Ito, O. and Watanabe, I. 1983. The relationship between combined nitrogen uptakes and nitrogen fixation in Azolla-Anabaena symbiosis. *New Phytol.,* 95: 647-54.

Jackson, M.L. 1958. *Soil Chemical Analysis.* Prentice hall of India, New Delhi. p.498.

Jana, B.B.; Lachmi and Chakarabarti, L. 1997. Effect of manuring rate on situ production of zooplankton Daphnia carinata. Aquaculture. 156 (1&2): 85-99.

Jhingran, A.G. 1988. Reservoir fisheries in India. *J. Ind. Fish. Assoc.* 18: 261-273.

Jhingran, A.G. 1989. *Reservoir Fisheries Management in India.* CICFRI Bull. 45: 65.

Jhingran, V.G. 1986. Aquaculture of Indian major carps. *In:* Billard, R. and Marcel, J. (eds.). *Aquaculture of Cyprinds.* INRA, Paris. pp. 335-346.

Jhingran, V.G. 1991. Fish and fisheries of India, Hindustan Publishing corpn. (India) New Delhi. pp. 511-516.

Jhingran, V.G. and Sharma, V.K. 1986. Integrated livestock-fish farming in India. In: *Research plans* for Integ. Aqua. FAO/UNDP Bangladesh, pp. 46-53.

Jiang, Chenzhong.; Cai, Shaozhu and Li Bingzheng, 1993. On the synthetical development of the Dongping Lake region. *Trans. Oceand Limnol,* 1: 60-63.

John, C.K. (1986). Personal communications (April-June). The Rubber Research Institute of Malaysia, PO Box 15, Kuala Lampur 01-02, Malaysia.

John, C.K. 1984. Treatment of Agri-industrial wastes using water hyacinth. *Water Sc. Tech.* 17: 781-790.

Jone, C. 1990. Water quality management. Proc. of the seminar-Farming the red clow fresh water cray fish, shelley. 21: pp. 8-12.

Kannaiyan, S. 1979. Utilization of *Azolla* for rice crop. *Fmr and Parliament.* 14 (8): 33-34.

Kant, R.; Pandey, S.D.; Sharma, S.K. and Rajnikant, 1996. Role of biological agents for control of mosquito breeding in rice fields. *Indian Journal of Malariology*. 33: 4, 209-215.

Kapur, K. 1981. The utilization of some organic wasters for fish culture. *Acta Hydrobiol.* 23: 95-102.

Kapur, K. 1984a. Effect of organic wastes on the primary and secondary levels of production in the treated waters. National Seminar on organic waste Utilization and Vermicomposting. Sambalpur, Dec. 1984, AVP-11.

Kapur, K. 1984b. The economic aspects of utilization of livestock wastes for carp culture. National seminar on organic waste utilization and vermicompositing, Sambalpur, Dec. 1984., AVP-12.

Kapur, K. and Lal, K.K. 1986a. Relative potency of certain livestock wastes for fish culture. In: Mclean. J. (ed.) *Proc. First Asian Fisheries Forum*, Manilla, Philippines.

Kapur, K. and Lal, K.K. 1986b.The chemical quality of wastes treated waters and its relation with patterns of Zooplankton population. In: Melean, J. (ed.), Proc. First Asian Fisheries Forum, Manilla, Philippines.

Karause, H.R. 1962. Investigation of the decomposition of organic matter in natural waters. *FAO Fish Biol. Rep.* No. 34.

Kasmi and Trivedi, 1978. *Indian Pharmac.* Ind. 13(3): 21.

Kaunhog, M. 1996. Aquaculture sustainability in fish-livestock- crop integrated management. *Fishing Chimes.* 15(10): 9-11.

Kestemont, P. 1995. Different systems of carp production and their impacts on the environment. *Aquaculture.* 129: 347-372.

Khan, A.A. and Siddiqui, A. 1971. Primary production in a tropical fish pond at Aligarh, India. *Hydrobiologia*. 37: 447-456.

Khan, M.M. 1983. A *Primer on Azolla Production and Utilization in Agriculture*. University of the Philippines at Los Banhos, Philippine Council for Agricultureal Research and Resources Development (PCARRD) and Southest Asian Regional Center for Graduate Study and Research in Agriculture (SEARCA), Los Banos Philippines.

Khanna, S.S. 1993. *An introduction to fishes*. Central Book Depot, Allahabad. p. 530.

*Kickurth, R. 1975. Degradation and incorporation of nutrients from rural waste waters by plant *rhizosphere unde limnic conditions. Utilization of manure by land spreading*. Coordination of Agricultural Research, Commission of the European Communities.

Kiss, L. and Pekar, F. 1998. Effective manure production of fish pond raised ducks. *In:* Mathias, J.R.; Charles, A.T. and Baotong, H. (eds.), *Proc. Work. Integre. Fish Farming*, pp. 157-162.

Kok, T. 1974. *De reininging van advalwater van een kampeerterrain met behulp van een biezenveld*. H_2O (7): 537-44.

Krom, M.D. and Neori, A. 1989. A total nutrient budget for an experimental fish pond with circulatory moving seawater. *Aquaculture*. 15: 185-193,

Krom, M.D.; Porter, C. and Gordin, H. 1985. Nutrient budget of a marine fish pond in Eilat, Israel. *Aquaculture*. 51: 65-80.

Kumar, D. 2004. Study on efficacy of vermicompost as organic manure and earthworm meal as protein rich supplement in feeding of *Labeo rohita*. M.F.Sc. Thesis submitted to G.B.P.U.A.&T., Pantnagar.

Kumar, K.; Sehgal, K.L. and Sundar, S. 1979. Length-weight relationship and ponderal index of brown trout, *Salmo trutta fario* (Lin.) catches in the streams of Kashmir. *J. Fisheries Soc.* India. (1): 56–61.

Lakshmanan, M.A.V.; Sukumaran, K.K.; Murty, D.S.; Chakraborty, D.P. and Philipose, M.T. 1971. Preliminary observations on intensive fish farming in fresh water ponds by the composite culture of Indian and exotic species. *J. Inland Fish Soc. India*. 3: 1-21.

Lal, K. K. 1986. Toxicity and utilization of livestock wastes for zoo plankton and fish production. M.Sc. thesis, Haryana Agricultural University, Hisar.

Lal, K.K. and Kapur, K. 1987. Temperature versus biochemical oxygen demand its significance in quantitative and qualitative evaluation of certain livestock wastes for pisciculture. Acta Hydrochem.

Le Cren, E.D. 1951. The length-weight relationship and seasonal cycle in gonadal weight and condition in the perch (*Perca flaviatilis*). *J. Anim. Ecol.* 20: 201–219.

Lewis, W.M. Jr. 1979. Zooplankton community analysis: Studies on a tropical lake. Springer-Verlag, NewYork p.163.

Lin, H.R. 1982. Poly cultural system of fresh water fish in China. *Can. J. Fish. Aquat. Sci.* 39: 143-150.

Little, D. and Muir, J. 1987. A guide to integrated warm water Aquaculture. Institute of Aquaculture publications. Uni. of Sterling, Scotland, p. 238.

Liu chung Chu, 1979. Use of *Azolla* in rice production in China. In *Nitrogen and Rice*, pp. 375-394.

Liu, C. 1979. Use of Azolla in rice production in China. In: Nitrogen and Rice, International Rice Research Institute, Los Banos Laguna, Phillipines, pp. 374-375.

Liu, C. 1987a. Re-evaluation of Azolla utilization agricultural production. In *Azolla Utilization,* Proceeding of the workshop on Azolla use. International Rice Research Institute Los Banos, Laguna, Philippines. pp. 67-76.

Lovshin, L.L., N.B. Schwartz, V.G. de Castillo, C.R.Engle and U.L. Hatch (1986). Cooperatively managed rural pananian fish ponds: The integrated approach. International center for Aquaculure Research and Development, series M03, December,

Lumpkin, T.A. 1977. *Azolla* in Kwangtung Province. Peopleds Republic of China. *Int Rice Res. Newslett.* 2: 18.

Lumpkin, T.A. and Plucknett, D.L. 1980. *Azolla:* Botany Physiology and use as a green manure. *Econ. Bot.* 35(2): 111-153.

Lumpkin, T.A. and Plucknett, D.L. 1982. *Azolla* as a green manure, use and management in crop production. Westview Press; Bouldercolorado, USA. Series No. 15, p. 230.

Lyubumova, T.S. 1974. Effect of fertilization on Zooplankton growth in carp nursery ponds in Southern Urals. *Hydrobio, J.*

Magizharasan, M. and Indra. 1997. Limnological studies of a pond at Thillaividangan in South Arcot with reference to physico-hemical factors and Plankton. *Environment and Ecology.* 15(3): 692-695.

Malhotra, S. K. and Chauhan, R. S. 1984. Bionomics of hill stream cyprinids. VI. Length-weight relationship of *Labeo dero* (Ham) from India. Proc. of the Indian Academy of Sciences (Animal Science). 93 (5): 411–417.

Malhotra, S.K. 1982. Bionomics of hill stream syprinids III. Food, parasites and length weight relationship of garwal mahseer, *Tor tor* (Ham.); *Proc. Indian Acad. Sci.* (*Ani. Sc.*) 91: 479-485.

Malhotra, S.K. 1985. Bionomics of hill strean cyprinids I. Food, parasites and length weight relationship of *Labeo dyochellus* (Mc Clell). *Proc. Indian Acad. Sc.* (*Ani. Sc.*) 94(4): 377-381.

Malik, S. 1996. Assessment of Toxicity and utilization of Piggery waste as fed and fertilizer for fish production. Ph. D. thesis submitted to H.A.U. Hissar.

Mandal, B.K. and Ghosh, M.R. 1990. Bhasamam Jalaj Udvid (Aquatic Floating Plants). Paschimbanga Rajya Pustak Parsad, Calcutta.

Mandal, B.K. and Moitra, S.K. 1975. Studies on the bottom fauna of a fresh water fish pond at Burdwan. *J. Inland Fish. Soc. India.* 7: 43-48.

Manimaran, B.; Ramadhas, V. and Santhanam, R. 1997. Cultivation of duck weed lemna minor utilizing waste nutrients. Proc. of National seminar o n water quality in Aquaculture system, T.N. Vety. Ani. Sc. University: pp. 65-72.

Marco, V.; Acosta-Nassar; Julio, M.M. and Corredor, J.E. 1994. The nitrogen budget of a tropical semi-intensive fresh water fish culture pond. *J. World Aqua. Soc.* 25(2): 261-270.

Maricel Becerra, E. Murgueitio, G. Reyes and T.R. Preston. 1990. Azolla filiculoides as partial replacement for traditional protein supplements in diets for growing fattening pigs based on sugar cane juice. Livestock Research for Rural Development. Vol. 2 pp. 1-5.

Martin, J.L.M.; Yves, V.; Olivier, G. and Dominique, P. 1998. Shrimp rearing: stocking density, growth, impact on sediment, waste output and their relationship studied through the nitrogen budget in rearing ponds. *Aquaculture.* 164: 135-149.

McCombic, A.M. 1953. Factors influencing the growth of phytoplankton. *J. Fish Res.* B. Canada, 10(5): 253-282.

McIntire, C.D. and Boyd, C.E. 1962. Effects of artificial fertilization on plankton and benthos abundance in four experimental ponds. *Trans. Amer. Fish. Soc.* 191: 303-312.

Meno, A. 1988. Sugarcane juice as animal feed an overview, In: Sugarcane as feed, FAO, Rome pp. 153-163.

Merkens, J.C. and Downing, K.M. 1957. The effect of tension of dissolved oxygen on the toxicity of un-ionized ammonia to several species of fish. *Ann. Appl. Biol.* 45: 521-527.

Micha, J.C.; Antoine, T.; Wery, P. and Van Hove, C. 1988. Growth, ingestion capacity, comparative appetency and biochemical composition of *Oreochromis niloticus* and *Tilapia rendalli* fed with Azolla. In: Second International Symposium on Tilapia in Aquaculture (eds) R.S.V. Pullin, T. Bhukaswan, K. Tonguthai and J.L. machean), ICLARM Conference Proceedings, 15. Department of Fisheries, Bankok, Thailand and International Centre for Living Aquatic Resources Management, Manila, Philippines. pp. 347-55.

Milstein, A. 1990. Fish species interactions, *EIFAC/FAO Symposium on Production Enhancement in Still Water Pond Culture.* Prague, Czechoslovakia, EIFAC/90/Symp. R.5, p. 20.

Mishra, B.K.; Sahu, A.K and Pani, K.C. 1986. Recycling of the aquatic weed, water hyacinth and animal wastes in the rearing of IMC. *Aquaculture.* 68: 59-68.

Mitra, B.; Gupta, A. and Laha, U.K. 1987. Effects of some manures on the growth and production of major carps in village ponds of district Birbhum, West Bengal. *Environ. Ecol.* 5(2): 381-385.

Mohanty, A.N.; Chatterjee, D.K.; Saha, P.K. and Pani, K.C. 1994. Effect of varying C/N ratios on the mineralization of organic nitrogen in fish pond soils. *J. Aqua. Trop.* 9: 9-14.

Moore, A.W. 1969. Azolla biology and agronomic significance, *Bot. Rev.* 35: 17-35.

Mora, D. 1982. Sewdage Treatment in Hot climates. A wiley Inter Science Publication, Great Pritian, 168: 1-95.

Moyle, J.B. 1946. Some indices of lake productivity. *Trans. Ame. Fish. Soc.* 76: 322-334.

Mukherjee, B. and Pankajakshi, G.V.N. 1995. The impact of detergents on plankton diversity in fresh waters. *J. Enviorn. Biol.* 16(3): 211-218.

Mullar, Z. O. 1980. FAO- Animal production and health, paper no. 18: p. 78.

NACA.1989. Integrated fish farming in China. NACA. Tech. Manual 7. NACA Bangkok, Thailand, 278.

Nath, D.; Mandal, L.N.; Tripathi, S.D. and Karmakar, H.C. 1994. Correlation between fish production and physico-chemical parameters of water and soil in semi-intensive carp culture ponds of West Bengal, India. *J. Inland Fish. Soc. India.* 26(1): 106-115.

*Newell, R. 1980. The role of detritus in the nutrition of two marine deposit feeders *Proc. Zool. Soc.* London.

Odum, E.P. 1971. *Fundamental of Ecology.* W.B. Saunders Co., Philadelphia, New York, p. 574.

Oertli, B. 1995. Spatial and temporal distribution of the zoobenthos community in a wood land pond (Switzerland). *Hydrobiologia.* 300-301: 195-204.

Olah, J.; Sinha, V.R.P. Ayyappan, S. Porushothaaman, C.S. and Radheyshyam, S. 1986. Primary production and fish yields in fish ponds undes different Management Practices. Aquaculture, 58: 111-122.

Olsen, S.; Cole, C.V.; Watenabe, F.S. and Dean, L.A. 1954. Estimation of available phosphorus in soil by extraction with sodium by-carbonate. *Cire. U.S. Dept. Agric.* 939.

Oosten, V. J. and Eschneryer, H. Paul. 1956. Biology of young lake Trout (*Sacvellinus namaycush*) in lake Michigan. *Fish Wild Serv.* U.S. Dept. Int. 42: 48–55.

Oron, G. Wildschut, L.R. and Porath, D. (1984). Wastewater recycling by duckweed for protein production and effluent renovation. *Water Science and Technology*. 17: 803-817.

Ovie, S.I. and Adeniji, H.A. 1991. Zooplankton culture in outdoor concrete tanks; the effect of local fertilizer on zooplankton population development. *Ann. Rep. Natl. Inst. Freshwat. Fish. Res. Nigeria*, pp. 129-135.

Padmavathi, R.; Daniel Jameson, J.; Sanjeeviraj, G. and Ramadhas, V. 1997. Impact of nitrogen and phosphorus on brackish water pond fertility and effluent quality. *In:* Santhanam, R.; Ramadhas, V. and Gopalkrishnan, P. (eds.), *Proc. Nat. Sem. Water Quality Issues in Aqua. Systems*, pp. 83-89.

Pandey and malik, D.S. 2002. Study on zooplankton production with the application of pig dung biogas slurry. *J. Natcon* 14(2): 263-267.

Pandey, K.; Radhey, S.; Sant Prasad and Chaudhary, H.S. 1983. A study on the macro-zoobenthos and the physico-chemical characteristics of the bottom of Bakhira lake, U.P. India. *Int. Revue ges. Hydrobiol.* 69(4): 591-597.

Pant, M.C.; Gupta, P.K.; Pande, J.; Sharma, P.C. and Sharma, A.P. 1981. Certain aspects of water pollution in lake Nainital (U.P.) India. *Environm. Conserv.* 81: 113-117.

Parks, R.W.; Scarsbrook, E. and Boyd, C.E. 1975. Phytoplankton and Water Quality in a Fertilished Fish Pond. Auburn Univ. (Ala.) *Agr. Exp. Sta. Cir.* 224, p. 16.

Parnas, B. and Lahav, N. 1972. Movement of nitrates and other dissolved salts from a fish pond in to lake Kinneret, *Bamidgeh.* 24(4): 99-133.

Partohardjono, S.; Hendrick, V. and Bastaman, M. 1983. Effect of *Azolla* incorporation spacing and nitrogen fertilizer application on the growth and yield of wetland rice. *Conf. Centr. Res. Inst. Fd Crops.* Indonesia. 69: 11-21.

Pathak, V. 1981. Patterns of energy flow and nutrient cycle in integrated farming ecosystems. Lecture delivered in Summer Institute on "Farming System Integrating Agriculture Livestock and Fish culture" CIFRI, Barraackpore, W.B.

Pathani, S.S. and Das, S. M. 1980. A note on length-weight relationship and seasonal condition factor of Mahasheer, *Tor tor* and *T. putitora* Ham.). *J. Indian Fish Soc.* India. 12 (1): 140–143.

Patil, C.S. and Gouder, B.Y.M. 1985. Ecological study of fresh water zooplankton of a subtopical pond (Karnataka, India) *Int Rev. Gesmet. Hydrobiol.* 70(2): 259-267.

Patnaik, S.; Aayyappan, S.; Saha, P.K.; Jena, S. and Das, K.M. 1988. Plankton dynamics on fresh water nursery rearing and stocking ponds. Proceedings of First Fisheries Forum. 4-8 December, 1987. Manglore, Karnataka, pp. 17-20.

Patra, B.C. and Ray, A. K. 1988. Observation on the influence of three organic manures on the growth and yield of Indian major carps under pond culture. *J. Inland Fish Soc.* 20(2): 61-63.

Patra, S. and Santra, S.S. 1999. Bioremediation of waste water from mill using aquatic Macrophytes. Proc. National Conf. On Industry and Env. (Ed. By R.K. Trivedi) Karad. pp. 164-168.

Patrick, R. 1953. Aquatic organisms as aid in solving waste disposal problem. *Sewage Ind. Wastes.* 25: 210-214.

Pekar, F. and Olah, J. 1991. Carbon pathways, bioenergetic efficiencies and energy cost in fish-cum-livestock ecosystem. *In:* Pauw, N.D. and Jojee, J. (eds.), *Aquaculture and the Environment*, EAS Sp. Publ., Ghent, Belzium. 16: 71-78.

Pennak, R.W. 1957. Species composition of limnetic zooplankton communities. *Lomnol. and Oceanogr.* 2: 222-232.

Pennak, R.W. 1978. *Fresh water invertebrates of the United States.* A Willey inter science publication John Willey and Sons, New York. p. 803.

Peters, G.A. and Calvert, H.E. 1984. The *Azolla*-Anabaena *symbiosis. In* Advances in Agricultural Microbiology. (Ed.) N.S. Subba Rao. Oxford and IBH Publishing and Co., New Delhi. pp. 191-218.

Peters, G.A. Mayne, B.C. Ray, T.B. and Toia, R.E. Jr. 1979. Physiology and biochemistry of the Azolla-Anabaena symbiosis. In: Nigroten and Rice. International Rice Research Institute Los Banos, Laguna, Philippines. pp. 325-44.

Philipose, M.T. 1959. Freshwater phytoplankton of inland fisheries. *Proc. Symp. Algology,* ICAR, New Delhi, pp. 271-291.

Philipose, M.T. 1960. Fresh water phytoplankton of inland fisheries. In: Kechroo. P. (ed.) *Proceeding of the symposium on Algology*. ICAR. New Delhi, pp. 272-291.

Phillipose, M.T. 1967. *Chlorococcales*. ICAR, New Delhi, pp. 271-291.

Pillay, T.V.R. 1990. Aquaculture: Principles and Practices. Fishing News Books, London, pp. 526-542.

Press, C.M.; Mahaffee, W.F.; Edward, J.H. and Kloepper, J.W. 1996. Organic by-product effects on soil chemical properties and microbial communities. *Compost Science and Utilization*. 4(2): 70-80.

Prinsloo, J.F. and Schoonbee, H.J. (1987a). Investigations into the feasibility of a duck–fish–vegetable integrated agriculture-aquaculture system for developing areas in South Africa. Water SA 13(2): 109-118.

Prinsloo, J.F. and Schoonbee, H.J. 1986. Summer yield of fish in polyculture in Transkei, South Africa, using pig manure with and without formulated fee. S. AFR. J. ANIM. Sc., 16(2): 65-71.

Pronloo, J. F. and Schoonbee, H. J. (1987 b). The use of sheep manure as nutrient with fish feed in pond fish polyculture in Transkei. Water SA 13(2): 119-124.

Pushpendra and Madhyastha, M.N. 1994. Seasonal variation and diversity of zooplankton in a small pond near Mangalore. *J. Ecobiol.* 6(3): 197-200.

Qadri, M. Y. and Mir, S. 1980. Length-weight relationship of *O. plagiostromus* (McClell). Geobios. 71 (4): 158–159.

Quasim, S.Z. Wallershans, W. Bhattathiri, P.M.A. and Abidi, S.A.H. 1969. Organic production in tropical estuary. *Proc. Indian Acad. Sci.*, 69: 51-94.

Radhey S. 1995. Observation of fish yield and employment generation in relation to farm area and multi-commodity production in integrated system in China. *Fishing Chimes*. 15(7): 29-31.

Raghavan, S.L.; Rao, N.G.S. and Rahman, M.F. 1986. Hydroniological investigation in two perennial ponds of Anekal Taluk, Karnataka. *J. Inland fish Soc.* India. 18(2): 52-59.

Rana, G.C.; Sengupta, K.K. and Santra, S.C. 1996. Limnological characteristics of beels with reference to fish yield in tropics. *J. Inland Fish. Soc. India.* 28(1): 59-66.

Rao, K.J. and Raju, T.S.R. 1989. Observations on poly culture of carps in large fresh water ponds of kolleru lake. *J. Aqua. Trop.* 4: 157-164.

Rao, N.G.S.; David, A.; Raghavan, S.L. and Rahman, M.F. 1982. Observations on the limnology of a peninsular perennial tank. *Mysore J. Agric. Sci.* 16(1): 75-84.

Rao, N.S.S.; Venkalaraman, G.S. and Kannaiyan, S. 1993. Nitrogen fixation by Azolla and its contribution of nitrogen to the rice crop. In: Biological nitrogen fixation, ICAR–New Delhi: pp. 77-106.

Rawat, H.S. 1991. Studies on the limnology and fisheries of Tumaria reservoir at Kashipur (Nainital). Ph.D. Thesis, Kumaun, University, Nainitial. p. 188.

Ray, P. and David, A. 1966. Effects of industrial wastes and sewage upon the chemical and biological composition and fisheries of river Ganga at Kanpur (U.P.). *Environ. Hlth.,* 8: 307-339.

Reddy, K.R. and DeBusk, W.F. (1985). Growth characteristics of aquatic macrophytes cultured in nutrient-enriched water. II. Azolla, duckweed, and salvinia. *Economic Botany.* 39, p. 200.

Reid, G.K. 1961. *Ecology in Inland Waters and Estuaries*. Reinhold publishing crop, New York. p. 375.

Rengaraj, S.; Sivabalan, R.; Arabindoo, B. and Murugesan, V. 1999. Removal of Orthachlorophenol from Aqueous Solution by Activated Carbon Prepared from Rubber Seed Coat. Proceeding of National conference on Industry and Environment, Karad (ed. R.K. Trivedy). Daya Publishing House–Delhi. pp. 68-75.

Rodhe, W. 1969. Crystalization of eutrophication concept in North America. *In: Eutrophication, causes, consequences and Correctives,* National Academy of Sciences, Washington, D.C. 50-64.

Ruttner, F. 1953. *Fundamentals of limnology*. 3rd ed. University of Toronto Press. Toronto. p. 242.

Saha, L.C. 1985. Changes in the properties of bottom soil of two fresh water ponds in relation to ecological factors. *Indian J. Ecol.* 12(1): 147-150.

Saini, V.P. Sharma, S.K. and Sharma, L.L. 1992. Production of algae (*Calothrix* spp.) using urban sewage. *Poll. Res.* 11(1): 27-31.

Santiago, C.B. Aldaba, M.B. Reyers, O.S. and Laron, M.A. 1988. Response of Nile Tilapia (*Oreochromis nilotices*) fry to diets containing Azolla meal. In: The Second International Symposium for Tilapia in Aquaculture (eds R.S.V. Pullin, T. Bhukaswan, K. Tonguthai and JH.L. Maclean). ICLARM Conference Proceedings, 15. Department of Fisheries, Bankok, Thailand and International Center for Living Aquatic Resources Management, Manila, Philippines. pp. 377-382.

Sastry, V.K. and Malik, D.S. 1992. Seasonal variation in physico-chemical characteristics of a fresh water fish pond at village Baliana, Rohatak, Haryana. *Ad. Bios.* 11(1): 25-32.

Schroedar, G. L. 1980. Fish farming in manure loaded ponds. In: R. S. V. Pullin and Zaid, H. Shehadesh (Eds.) *Proc. ICLARM-SEARCA. Cont. Integ. Agri, Aqua. Fmg, Systems. Manila, Philippines.* pp. 73-86.

Schroeder, G. L.; Wohlfarth, G.; Alkan, A.; Halery, A. andKrueger, H. 1990. The dominance of algal based food webs in fish pond receiving chemical fertilizers plus organic manures. *Aquaculture,* 86 (2-3): 219-229.

Schroeder, G.L. 1974. Use of fluid cowshed manures in fish ponds. Bamidgeh, 26: 84-96.

Schroeder, G.L. 1975. Night tune material balance for oxygen in fish ponds. Bamidgeh, 7: 65-74.

Schroeder, G.L. 1977. Agricultural wastes in fish farming. A commercial application of the culture of single-celled organisms for protein production. Water Res. 11: 419-420.

Schroeder, G.L. 1987. Carbon and nitrogen budgets in manured fish ponds on Israel's Coastal plain. *Aquaculture.* 62: 259-279.

Seidel, K. 1976. Macrophytes and water purification. J. Tourbier and R.W. Pierson (Eds). *Biological control of water pollution.* University of Pennsylvania Press. p. 109.

Seidel, K.; Happel, H. and Graue, G. 1978. *Contributions to revialisation of water.* Stiftung Limnologische Arbeitsgruppe Dr. C.V. Seidal, meersburg, West Germany.

Sen, P.R.; Rao, N.G.S. Chakrabarty, R.D.; Kar, S.K. and Jena, S. 1978. Effect of additions of fertilizer of vegetation on growth of Indian major carps in ponds containing grass carp. *Porg. Fish. Cult.* 40(2): 69-70.

Shang, Y.C. and B.A. Costa- Pierce, 1983. Integrated Agriculture Aquaculture Farming Systems, Some economic aspects. *J. World Maricul. Soc.* 14: 523-530.

Shankar, K.M. 1978. Studies on the effect of fertilizing fish ponds by combing super-phosphate with cattle dung or sewage sludge cake. M.F. Sc. Thesis, University of Agricultural Sciences, Bangalore.

Sharma, A.P. 1980. Phytoplankton primary production and nutrient relations in Nainital lake. Ph.D. Thesis, Kumaun University, Nainital p. 317.

Sharma, A.P.; Singh, U.P.; Chauhan, R.S. and Singh, V.K. 1998. Duck-cum-fish culture in *Tarai* region of U.P. In: Ecological Agriculture and sustainable development. Vol. I. *Indian Eco. Soc.*, Ludhiana, pp. 258-292.

Sharma, B. K. Das, M. K. and Chakroberty, D. P. 1985. *Aquaculture Extension Manual New Series* No. 5, CIFRI, Barrackpore.

Sharma, B.K.; Das, M.K.; Das, S.R. and Kumar, Dilip. 1989. Operational research on composite fish culture. Paper presented at the annual workshop on operational research project, held at Sukhadia University, Udaipur, Rajasthan.

Sharma, B.K. and Das, M.K. 1988. Integrated fish-livestock-crop farming. In: Mohan Joseph (eds.). *Proc.* First Indian Fisheries forum, AFSIB, Manglore, pp. 27-29.

Sharma, B.K. and Olah, J. 1986. Integrated fish-pig farming in India and Hungary. *Aquaculture.* 54: 135-139.

Sharma, B.K.; Das, M.K. and Chakraborty, D.P. 1985. *Package of Practices for Increasing Production in Fish-cum-Livestock Farming System.* Aquaculture Extension Manual, CIFRI, Barrackpore. 5: p. 32.

Sharma, B.K.; Kumar, D.; Das, M.K. and Chakraborty, D.P. 1979. Observation on swine dung recycling in composite fish culture *Symp.* Inland Aquaculture, CIFRI, Barrackpore.

Sharma, O.P. and Saini, V.P. 1992. Evaluation of pig manure fertilization in relation to zooplankton production and water quality. *J. Eco. Biol.* 4(1): 27-31.

Sharma, P.C. and Pant, M.C. 1984. Abundance and community structure of limnetic zooplankters in Kumaun lakes, India. *Int. Revue Ges. Hydrobiol.* 69(1): 91-109.

Shiomi, N. and Kitoh, S. 1997. The growth nitrogen fixation and nutrient removal capacity of *Azolla* in polluted water. Res. Inst. for Advanced Sci and Technol. Vol. 2, no. 3-4, pp. 175-179.

Shroff, K.C. 1977. Recovery and reuse of materials from pollutants by water hyacinth. Paper at the seminar of all India manufacturers. R.D. Organisation, Bombay.

Shuter, B.J. and Ing, K.K. 1997. Factors affecting the production of zooplankton in lakes. *Can. J. Fish.* 54(2): 359-377.

Sin, A.W. 1980. Integrated animal fish husbandry system in Hong Kong with case studies on duck-fish and goose-fish systems. *In:* Pullin, R.S.V. and Shehadeh, Z.H. (eds.), *Integrated Agri. Aqua. Fmg. Systems. ICLARM Conf. Proc.* 4: 113-123.

Singh, A.K. and Singh, D.K. 1999. *J. of Environmental Bio.* 20(3): 263-270.

Singh, A.M. and Kumari, S. 1994. Seasonal standing stock of phytoplankton communities in a fish pond at Munger, Bihar. *Environment and Ecology.* 12(1): 236-238.

Singh, C.S. and Das, M. 1993. A self reliant and sustainable production system. *In:* Proceedings of National Seminar on, Green Revolution: Impact on Health and Environment, at G.B.P.U.A.&T., Pantnagar. 29-30, March: 8-11.

Singh, C.S.; Sharma, A.P. and Deorari, B.P. 1990. Analysis of plankton population in relation to fisheries in Nanak Sagar reservoir, Nainital. *Proc. Nat. Sem. Rec. Adv. Hydrobio.* 23-25 Oct. Devi Ahilya Univ.

Singh, G.P. and Singh, U. 1996. Seasonal variations of macrozoobenthos of Rajendra Sarovar, Chaapra, Bihar, India, *J. Environ. Biol.* 17(3): 205-209.

Singh, R.K. 1977. Use of *Azolla* plants as organic fertilizer and feed. *Indian Fmg.* 27: 19-22.

Singh, R.K. 1979. Soil character of Rihand reservoir. *In: Summer Institute on Capture and Culture Fisheries of the Man-made Lakes in India*. CIFRI, Barrackpore. 159-162.

Singh, S.B.; Sukumaran, K.K.; Chakrabarti, P.C. and Bagchi, M.M. 1972. Observations on composite culture of exotic carps. *J. Inland Fish.Soc. India*. 4: 38-50.

Singh, V.K. 2002. Nutrient dynamics in duck- fish integrated farming system. P. H. D. Thesis, Kumaun University, Nainital.

Singh, V.K. 1996. Comparative effect of different organic manure on fish growth and assessment of pond water on maize growth, M.F.Sc. Thesis G.B. Pant University of Agriculture and Technology, Pantnagar. p. 89.

Singh, V.K. and Sharma, A.P. 1998. Community structure of Plankton in fish ponds manured with there organic manures. *Him. J. Env. Zool.* Vol. 12, pp. 91-98.

Singh, V.K. and Sharma, A.P. 1998. Economic analysis of carp culture in Tarai region of U.P. State. *Fishing Chimes.* 18(4): 23-26.

Singh, V.K. and Sharma, A.P. 1999. Hydrological characteristics and primary production in fish ponds, manured with different organic manures. *Indian J. Fish.* 46(1): 79-85.

Singh, V.K.; Chauhan, R.S. and Singh, U.P. 2000. Effect of differential stocking and phased harvesting on the development of plankton and fish production in silted non-drainable ponds. *Aquacult.* 1(1): 43-49.

Sinha, S.V. and Sinha, L.P. 1969. Studies on use of water hyacinth culture in oxidation ponds treating digested sugar wastes and effluents of septic tank. *Indian J. Environ. Hlth.*,11: 197-207.

Sinha, V.R.P.; Gupta, M.V.; Banerjee, M.K. and Kumar, D. 1973. Composite fish culture at Kalyani, West Bengal. *J. Inland Fish. Soc. India*. 5: 201-207.

Sivakami, R.; Prem Kishore, G. and Chandran, M.R. 1995. Diel migratory pattern of plankton in Ayilapettai pond at Tiruchirappalli, India. *J. Freshwater Biol.* 7(2): 109-116.

Sounders, R.M.K. and Flower, K. 1992. A morphological taxonomic revision of Azolla section Phizospherma (Mey.), Mett. (Azollaceae) Bot. Jour. Lilnn. Sec. 109(3): 329-357.

Sreenivasan, A. 1964. Hydrobiological studiy of a tropical impoundments. Limnology of Bhawani sagar reservoir, Madras State India for the year 1959-61. *Hydrobiologia.* 24(4): 514-535.

Sreenivasan, A. 1967. Fish production in some rural demonstration ponds in Madras (India) with an account of the chemistry of water and soil. *FAO Fish. Rep.* 44(3): 179-197.

Sreenivasan, A. 1976. Limnology of an upland lake. *Madras J. Fish.* 6: 9-14.

Srisumantach, V.; Soungchompan, R.; Sansrimahachal, C. 1982. Impact of manure on the ecology of fish pond. *Thai. Fish. Gaz.,* 35(1): 101-108.

Stephens, D.W. and Gillespie, D.M. 1976. Phytoplankton production in the great salt lake Utah and a laboratory study of algal response to enrichment. *Limnol. Oceanogr.* 21: 74-85.

Stewart, B.A.; Blake, J.; Donald, J. and Magette, W. 1992. Effect of animal manure on soil physical and chemical properties. Proceedings of the National Workshop, 29-31 July, 1991. Kansan city, Missouri, U.S.A, pp. 149-154.

Sukhtme, P.V. 1975. Human protein needs and the relation role of energy and protein in meeting them. In: F. Steele and A. Bourne (Editors), The Man/Food Equation, Academic Press, London, New York and san Francisco, pp. 53-75.

Swarnalatha, Paka and Rao, A.N. 1997. Inter-relationship of Physico-cheimcal factors of a pond. *J. Environ. Biol.* 18(1): 67-72.

Swatdee, P. and Seetanum, W. 1979. *Azolla* as nitrogen source for rice in northeast Thailand. *Int. Rice. Res. Newslett.* 4(5): 24.

Swingle, H.S. 1957. Relationship of pH of pond water to their suitability for fish culture. *9th Pacific. Sci. Congr., Bankok-Thailand* (cited by C.F. Hickling, 1962).

Tan, D.S.P. and K.H. Khoo. 1980. The integration of fish farming with agriculture faing system, R.S.V. Pullin and Z.H. Shehadeh (eds.) ICLARM conf. Proc. 4: 175-187.

Thakur, N.K. and Das, M. K. 1974. Length-weight relationship of *H. fossiles* (Block). *J. Inland Fish. Soc.* India. 6: 95–96.

Timm, T.; Kangur, K.; Timm, H.; Timm, V.; Timm, T. and Noges, T. 1996. Macrozoobenthos of lake Pepsi-Pihkva: Taxonomical composition, obundance, biomass, *Hydrobiologia*. 338(1-3): 139-154.

Tonapi, G.T. 1980. *Fresh water animals of India* (An ecological approach) by Mohan Primlani. Oxford and IBH Publishing Co., New Delhi.

Tripathi, C.K.M. 1982. Investigation on Ganga river to determine biological indicators of water quality. Ph.D. Thesis, Banaras Hindu University, Varanasi.

Trivedi, R.K. and Khourane, B.V. 1985. Aabsorption of nutrients from industrial waste water by water hyacinth. *Comm Physiol. Ecol.* 10: 123-128.

Tuan, D.T. and Thuyet, T. 1979. Use of *Azolla* in rice production in Vietnam. In Nitrogen and Rice. International Rice Research Institute, Philippines. pp. 359-405

Tung, H.F. and Watanabe, I. 1983. Differential response of *Azolla-Anabaena* associations to high temperature and minus phosphorus treatments. *New Phytol.*, 93(3): 423-31.

Uheda, E.; Kitoh, S. and Shiomi, N. 1997. Response of siz azolla species to transient high temperature stress. Aquatic-Botany. Vol. 64, No. 1, pp. 87-92.

Van Hove C. and Lopez, Y. 1983. Fisiologia de Azolla. In workshop on assessment of Azolla use in Tropical Latin America Chidago; Peru (Mimecografo).

Van Hove, C. 1989. Azolla and its multiple uses with emphasis of Africa, FAO, Rome, p. 52.

Van Hove, C. and Lejeune, A. 2002. The Azolla and beena symbiosis. Biology and Environment: Proc. of the Royal Irish Academy, Vol. 102B No. p. 23-26.

Varadi, L. 1999. Integrated animal husbandry. In: Proc. FAO-EIFAC *Symp.* (15-18 May, 90). Research Institute of fish culture and Hydrology, Vodnany, Czechoslovakia, pp. 123-128.

Vasisht, H.S. and Shekhar, C. 1983. Ecology of waste waters of the post-graduate University of Medical Education and Research and Punjab University Campus, Chandigarh. *Indian J. Ecol.* 10: 1-6.

Verma, S.R. and Shukla, G.R. 1969. Pollution in a perennial stream "Khala" by the sugar factory effluent near Laksar (District Saharanpur), U.P. India. *Ind. J. Environ. Health.* 11: 145-160.

Vijverberg, J.; Arunachalam, M. and Amarasinghe, P.B. 1987. Seasonal dynamics of copepods and cladocerans in five Sri Lanka reservoirs. *Proc. Work. Reservoir Fish. Mang. Deve. Asia,* Nepal, IDRC. 264(C): 58-68.

Vincke, M. and U.W. Schmidt. 1979. Report of a mission to identify opportunity for small scale fish farming. Aquacultaure development in the provinces of loja, El. Oro and Zamora–Chinchipe in Ecuador, FAO Rome, p. 38.

Vincke, M.M.J. 1991. Integrated farming of fish livestock: Present status and future development. *In: Proc. FAO/IPT Int. Work. Integra. Livestock Fish Prod. Systems.* Kualalumpur, Malaysia, pp. 27-34.

Vitenhimer, M.I.L.; Lopes, P.V.L.; Lanzer, R.M.; Pones, D.M.C. and Silva, M.C.P. 1990. The marginal zoobenthos of a southern Brazilian dam. Rev. *Braassilerira* de Bio. 50(1): 7-14.

Vollenweider, R.A. 1968. *Scientific fundamentals of the eutrophication of lakes and flowing waters with particular reference to nitrogen and phosphorus as factors in eutrophication.* Organization for Economic Co-operation and Development, Directorate of Scientific Affairs, Paris, Report. p. 220.

Wang, T.C. and Fullar, M.F. 1989. The optimum dietary amino acid pattern for growing pigs, experiment by amino acid deletion, *British Journal of Nutrition.* 62: 77-89.

Ward, H.B. and Whipple, G.C. 1959. *Freshwater biology.* Edmontson, W.T. (ed), 2nd Edition, New York.

Watanabe, I. 1977. *Azolla* utilization in rice culture. *Nt Rice Res, Neslett.* 2: 3.

Watanabe, I. and Furusaka, C. 1980. Microbial ecology of flooded soils. *Adv. Microbiol. Ecol.* 4: 125-168.

Watanabe, I. Berja, N.S. and Rosario, D.C. 1980. Growth of *Azolla* in paddy field as affected by phosphorus fertilizer, *Soil Sci. PI. Nutr.* 26(3): 301-307.

Weisner, S.E.B. Eriksson, P.G. Graneli, W. and Leonardson, L. (1994). Influence of macrophytes on nitrogen removal in wetlands. *Ambio.* 23(6): 363-433.

Welch, P.S. 1948. *Limnological methods.* McGraw Hill Co., New York. p. 381.

Welch, P.S. 1952. *Limnology.* New York, Toronto and London McGraw Hill Book Co., 2nd Edition. p. 538.

Westlake, D.E. 1980. Primary production. *In:* Le Cren, E.D.; Lowe, M.C. and Connell, R.H. (eds.). The functioning of fresh water ecosystems. Cambridge Univ. Press, Cambridge.

Westlake, D.F. 1963. Comparisons of plant productivity. *Biol. Rev.* 38: 385-425.

Wetzel, R.G. 1966. Variations in productivity of Goose and hypertrophic Sylvan lakes, Indiana. *Invest. Indiana lakes and Streams.* 7: 147-184.

Wetzel, R.G. 1975. *Limnology.* W.B. Saunders Co., Philadelphia, U.S.A. p. 743.

Wilber, C.G. 1971. The biological aspects of Water Pollution. Charles Thomas Publishers, U.S.A. p. 266.

Wohlfarth, G.W. and Schroeder, G.L. 1979. Use of manure in fish farming-A review. *Agril. Wastes.* 1(4): 279-299.

Wohlfarth, G.W. and Schroeder, G.L. 1987. Use of manure in fish farming, a review, Agriculture Waste. 1 (4): 279- 299.

Wolverton, B.C. and McDonald, R.C. 1975a. *Water hyacinths and alligator weeds for removal of lead and mercury from polluted waters.* National Space Technology Lab, Bayt St. Louis, MI. NASA Tech. Memo No. TM-X-72723.

Wolverton, B.C. and McDonald, R.C. 1975b. *Water hyacinths and alligaytor weeds for removal of silver, cabalt and strontium from polluted waters.* NASA Tech. Memo No. TM-X-72723.

Worsham, R.L. 1975. Nitrogen and phosphorus levels in water associated with a channel cat fish feeding operation. *Trans. Am. Fish. Soc.* 104(4): 811-815.

Woyanarovich, E. 1980. Utilization of Piggery waste in fish ponds. In: R.S.V. Pullin and Zald, H. Shehadeh (eds.). Proc. ICLARM-

SEARCA. *Conf.* On Inte. Agri. Aqua. Fmg. Systems, Manila, Philippines. pp. 125-128.

Woynarovich, E. 1979. The feasibility of combining animal husbandry with fish farming, with special reference to duck and pig production. *In:* Pillay, T.V.R. and Dill, W.A. (eds.), *Advances in Aquaculture*. Fishing News Books, Oxford. pp. 203-208.

Wright, J.C. 1959. Limnology of Canyon Ferry reservoir II. phytoplankton standing crop and primary production. *Limnol. Oceanogr.* 4: 235-245.

Yadav, N.K. 1987. Utilization of Domestic Sewage and Livestock wastes for craps Fish Culture, Ph.D. Thesis, Harayana Agricultural University, Hisar.

Yadav, S.B. 1979. A portable dissolved oxygen analyzer for the fish farming Industry Bamidgeh. Vol. 31, No. 3, p. 69.

Yadava, N.K. 1988. Pre-stocking Management Practices in Nursery Fish ponds. *Haryana Farming*. June, 1988.

Yadava, N.K. and Kapil, R.P. 1990. Utilization of polluted sewage water for Hisar (Haryana) for aquaculture. *Proc. Environ. Pollution.* pp. 153-160.

Yang, Yijing.; Zhu, Jun.; Huan, Dan. And Wan, Junhua. 1987. Studies on the effects of fresh and fermented manure on fish production. Bangkok Thailand, NACA: p. 8.

Yeoh, B.CG. (1993). Use of water hyacinth (*Eichhornia crassipes)* in upgrading small agro-industrial waste water treatment plant. *Water Science and Technology.* 28: 207-213.

Zhang, Zhuangta, Ke Yu-si, Ling De-Quan, Duang Bing-Yuan and Liu Xi-Lian. 1987. Utilization of Azolla in agriculture production in Guangdong Province, China. In: *Azolla Utilization,* Proceedings of the Workshop on Azolla use. International Rice Research Institute. Los Banos, Laguna, Philippines. pp. 141-145.

Zheng, W.W.; Nilsson, M.; Bergman, B.; and Rasmussen, U. 1999. Genetic diversity and classification of cyanobacteria in different Azolla species by the use of PCR fingerprinting. Theoretical and Applied Genetics. Vol. 99, no. 7/8, pp. 1187-1193.

Zhu, Yun.; Yang, Yejin.; Wan, Junhua. and Hua, Dan. 1987. Primary studies on frequencies of fertilization of fish ponds. Bongkok, Thailand, NACA, p 11.

Zhu, Yun; Ye, J.; Wan Jan. H. and Hua, Dan. 1990. Priliminary studies on the effects of frequencies of manure application on fish yield. Bangkok, Thailand, NACA, p 1.

Zoccarato, I.; Benatti, G.; Calvi, S.L. and Bianchini, M.L. 1995. Use of pig manure as fertilizer with and without supplementary feed in pond carp production in Northern Italy. *Aquaculture*. 129: 387-390.

Zoccarato, M.L. 1995. Use of pig manure as fertilizer with and out supplementary feed in fish pond carp production in Northern Italy. *Aquaculture*.129: 387-390.

Index